A Christian's Guide to Refuting Evolution

What Students Aren't Being Told

A Christian's Guide to Refuting Evolution
What students aren't being told

Copyright October 2024

Author: Bruce Malone
Contributing Editor: Julie Von Vett
Cover and Layout Design: Annelise Smith
Images used by permission of Freepik.com, istock.com, pixabay.com, Pexels.com and google.com/images

978-1-939456-41-0

Printed in China
First printing – October 2024

All Scripture quotations are taken from the KJV Bible unless otherwise indicated

Permission is granted to copy and distribute materials in this book for nonprofit educational Purposes without prior written consent from the author or publisher. Permission to republish any part of this book in any format, printed or electronic, for any other purpose is reserved.

Published by: Search for the Truth Publications
    www.searchforthetruth.net

For Additional Information:
Searchforthetruth.net
Creation101.org
Creation.com
ICR.org
Answersingenesis.com

TABLE OF CONTENTS

Section I - Why the Topic of Origins Matters
    A) Purpose Of This Guidebook
    B) What Is Creation? What Is Evolution?
    C) Two Models For Origin – Evolution And Creation
    D) Why Is Evolution So Widely Accepted?

Section II – Problems with Chemical Evolution
    A) Could Chemicals Come Alive?
    B) Problems With Origin Of Life Experiments

Section III – Problems with Biological Evolution
    A) The Classifying Of Life
        a) The Linnaeus System Of Classification
        b) Do Evolutionary Trees Prove Evolutionary Relationship?
        c) Do Similar (Homologous) Structures Show Evolutionary Relationship?
        d) Have Dinosaurs Turned Into Birds?
        e) Is There 98% Similarity Between Human And Chimpanzee Dna?

    B) Does The Fossil Record Prove Evolution?
        a) Have Fossils Formed Over Millions Of Years?
        b) Does The Cambrian Explosion Support Creation Or Evolution?
        c) Does The Sequence Of Fossils In The Rock Layers Prove Evolution?
        d) Does Pangaea (Tectonic Plate Movement) Prove The Evolutionary Time-Frame?
        e) Living Fossils
        f) Has The Grand Canyon Formed Over Millions Of Years?
        g) Was Dinosaur Extinction Caused By A Meteorite Impact?

Section IV – Problems with the Mechanisms of Evolution
    A) Natural Selection
        a) Darwin's Finches
        b) Dog Variation
        c) Peppered Moths
        d) Drug Resistant Bacteria
    B) Beneficial Mutations
        a) Sickle-cell Anemia
        b) Blind Cave Creatures
        c) Lenski's Bacteria
    C) Common Evolution "Proofs"
        a) Vestigial Organs
        b) Junk DNA
        c) Embryo Similarity
        d) Fossil Evolution
            i) Horses
            ii) Whales
    D) Human Evolution

Section V – Problems with Cosmic Evolution
    a) Can Nothing Turn Into Everything?
    b) Can The Big Bang Explosion Create Order?
    c) Can The Universe Explosively Expand?
    d) Can Gas Condense To Become A Star?

Section VI – Age of Creation Issues
    a) How Dating Methods Work
    b) Radiocarbon Dating
    c) Radiometric Dating Used for Igneous Rocks
    d) Recent Creation Evidence

Section VII – Conclusion – Why it Matters?
    a) Educational And Scientific Issues
    b) Biblical Issues

SECTION I - WHY THE TOPIC OF ORIGINS MATTERS

A. Purpose of this Guidebook

Most people have seen or heard about evolution either in their college training or through popular media. Science teachers in particular have seen many textbook examples of how evolution purportedly explains the development of life here on Earth. God is never mentioned or considered when textbook curriculum talks about the evolution of life here on Earth. Yet, there are two possible explanations for the origin of all life and the entire universe - creation or evolution. But only one of these explanations - evolution (and the supposed evidence supporting this explanation) is increasingly promoted in school curriculum and textbooks.

The purpose of this guidebook is to give teachers and parents the information needed to allow children and friends to come to an informed decision of which of the two possibilities for our origin is actually the truth.[1] Because evolution is promoted in so many places by so much information (textbooks, magazines, movies, the Internet), the worst thing Christians can do is to ignore the subject. This guarantees that eventually, the majority of people will simply come to believe it is true, not because there is evidence supporting evolution, but simply because the evidence for the alternative, creation, is not seen.

The information in this guidebook is suitable for students from upper elementary through adults. Remember that children in grade school are already being exposed to evolutionary concepts which contradict the Bible in movies and on the Internet. You will need to modify the information contained in this book as necessary to fit the age of the student.

Nothing helps information to be remembered and understood better than student interaction and interesting demonstrations. Throughout this guidebook you will find highlighted demonstrations to drive a point home. These can be modified to use as home school activities or involve an entire classroom of students. Use your creativity to modify these demonstrations/activities in the best possible way with your children. The vast majority of Christian children are being indoctrinated with evolutionary principles in public schools. Look in their textbooks and do some of the activities provided in this guidebook, or have them watch some of the book's video QR links to give them a more balanced biblical perspective.

In summary, this guidebook is much more than just a book of information on the evidence for creation. Four things make this a valuable resource:

Throughout, you will find footnotes referencing more supporting information. Whenever possible, I have included links to Online papers pertaining to the various topics covered. Creation Ministry International (www.creation.com) has an excellent search engine on their website with over 13,000 articles and videos available - covering every conceivable aspect of the scientific evidence supporting creation. You will find these links in the reference section. As you prepare to cover each topic, these are an excellent source of additional information, enabling you to give a more informed presentation.

Wherever possible, I have added a direct link to videos, giving much more information on a given topic. These videos range from minute-long videos emphasizing a specific point to 45 minute teachings on the subject. You will need to preview these videos to determine if they are suitable for use with your children.

You will also find demonstrations and student activities throughout the book to allow children to see for themselves how the laws of science support creation and show the impossibility of evolutionary concepts. Doing an activity to learn a scientific principle is far more powerful than just being told that principle. Students rapidly forget something they have been told (especially as the media, textbooks, and other authority figures tell them the opposite), but they seldom forget something which they have experienced.

Each section ends with the Key Points to Remember. This makes it easy to summarize and helps solidify knowledge. Repetition is the key to long-term proficiency.

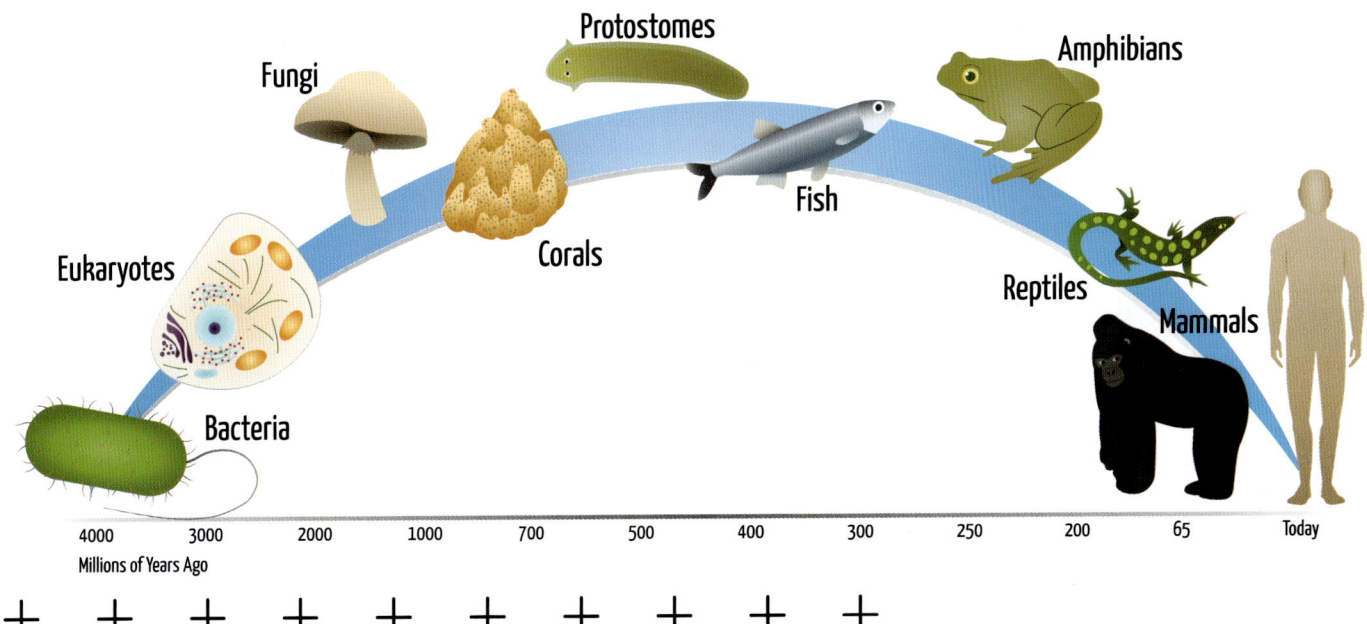

The chart is typical of how evolution is promoted. It shows a single type of bacteria coming alive on Earth about 3.5 billion years ago and diversifying into thousands of different kinds of bacteria today. Meanwhile, other bacteria turned into everything from whales to people over the same period of time. This guidebook will discuss WHY this is believed to be true. Meanwhile, let's continue to examine how the word "evolution" is used.

Common culture uses the word evolution to simply mean "beneficial change." When used in this way, evolution is true. Things can occasionally change in a beneficial way, but keep in mind that this happens because of programming within an organism - so these minor changes within an organism are ultimately guided by intelligence. We will discuss some of these examples of "evolution in action" later in this guidebook. But looking at the broader picture, things never get better or increase in complexity all by themselves - unless already programmed to do so or guided by intelligent effort. For instance, if left alone, buildings, cars, and even our bodies eventually deteriorate. Explosions never result in an increase in order. A scientific principle, called the second law of thermodynamics, explains why this is a fact of science. Everything always eventually winds down and deteriorates. There is no exception to this law of science, yet in biology it is assumed, not proven, that life has improved in complexity and ability all by itself.

But you might ask yourself, don't humans improve with time such as healing from illnesses and injuries, even if we do not go to a doctor? This does not happen by chance but is the result of mechanisms programmed into our cells so that our bodies can repair themselves and fight off diseases. Thus, even when evolution is used to mean beneficial change, we inherently know this change is guided by intelligence. It does not happen by itself.

Textbooks are filled with examples of evolution which show how life is programmed to change, adapt, and diversify. If this were the only way the word evolution was used, it would be absolutely true. It is often stated that "evolution is a scientific fact." When this statement is made, the evidence shown to students is always evidence that shows how one form of life can adapt, change, and diversify into similar, but slightly different, forms. This ability has been created or programmed into life. But then the same word, evolution, is used to mean something completely different – that one form of life can change into a completely different, more complex form of life, i.e. bacteria to people. In other words, more advance programming can appear on the DNA code by random change over time. This type of evolution, which has never been proven or observed by any scientific experiment or observation, is taken on faith. Evolution of bacteria to people is a faith that is merely assumed to be true. This guidebook will examine in detail the various purported "evidences" for evolution which are found in textbooks - so that you can help your students separate faith from scientific observation.

B. What is Creation? What is Evolution?

Creation is the belief that everything that exists did not make itself, but had a Creator who exists outside of His creation. Time, space, matter, and the very different classifications of biological life, could never have come into existence by themselves. Scientists who believe this use observations from science to support their belief.

The only alternative to creation is the belief that the universe came from nothing; chemicals came alive here on Earth billions of years ago; and bacteria slowly transformed into people by a process called "evolution." Evolution is such a common word that it is used everywhere from advertising slogans to political speeches to science discussions. It has come to mean everything from "change" to "the transformation of bacteria to people." When a word becomes so common that it is used in multiple ways for multiple purposes, it is often accepted as true no matter how it is used. Because of this widespread and common usage, many students simply accept it as true however the word is used. Sometimes the word "evolution" is used in a way that is purportedly supported by scientific observations and facts... but other times the same word is actually a philosophical belief - no different than a religious belief.[2] Your job as a teacher is to help students know the difference between facts, interpretations of facts, and dogmatically held beliefs. In other words, help students learn how to use critical thinking to determine what is the truth. This skill will serve them well throughout life - keeping them from being deceived when discussing a multitude of contentious issues.

The broadest scientific understanding of the word, evolution, is that of a process that believes all of life has arisen from a common source and are all related. This is commonly taught in biology as the "common descent of life" and assumed to be a fact of science. In simple language, evolution is the mechanism by which non-living chemicals turned into living bacteria, something as simple as a bacterium increased in complexity, and over enormous periods of time transformed into the amazing diversity of life and eventually turned into human beings.

CREATION ✝ ✝ ✝ ✝ ✝ ✝ ✝ ✝ ✝ ✝ ✝ ✝ ✝

The first intervention of God was the creation of the entire universe, fully formed and fully functional, including the wide variety of life and mankind, in six literal days. Exodus 20:11 states, "For in six days the LORD made heaven and earth, the sea, and all that in them is, and rested the seventh day: wherefore the LORD blessed the Sabbath Day, and hallowed it." Jesus confirmed in Mark 10:6 that, "from the beginning of creation (NOT 13 billion years later) God made them (human beings) male and female." When this event is left out of a scientist's thinking, he is guaranteed to misinterpret cosmology, biology, and all of history.

CURSE ✝ ✝ ✝ ✝ ✝ ✝ ✝ ✝ ✝ ✝ ✝ ✝ ✝

The second intervention of God was to curse a perfect creation so that mankind (who chose to rebel against the Creator, i.e. sinning) would not live forever separated from God. If this event is denied as a real event of history, human nature, the past, and the cure for mankind's true problems, will be misunderstood. In a way, the sad history of the world is mankind's vain attempt to look for meaning, purpose, and redemption in all the wrong places because humanity has denied that we brought death and sickness upon ourselves. We can do nothing on our own to pay for our rebellion.

WORLDWIDE FLOOD ✝ ✝ ✝ ✝ ✝ ✝ ✝ ✝ ✝ ✝ ✝

The third major intervention of God was a global, world-resurfacing flood upon the Earth approximately 4,500 years ago. When this is denied, all of geology, anthropology, paleontology, and world history will be misinterpreted. It is not that scientists are stupid or even trying to deceive - they are simply leaving the truth of the past out of their thinking and therefore not even looking for the truth. Rather than accepting the best explanation for the rock layers (a world-covering flood of the past), they rely on naturalistic explanations to explain the past. These misinterpretations always require enormous time periods, leave God and what He has told us out of their thinking, This results in belief in cosmic, chemical, biological, and geological evolution beliefs..

Throughout this guidebook we will discuss the common evidence proposed for evolution (which leaves God's three major interventions out), but also include evidence showing that these major interventions of God (creation, the curse, and the worldwide flood) are real historical events of the past. You will learn how the "evidence" being presented by evolutionists to support a molecules-to-man type of evolution, is misleading. Remember there are really only two possible explanations for our existence, and they cannot both be true:

Why Origins Matter

1. Natural laws of nature created the entire universe over billions of years and chemicals came alive to form bacteria which slowly turned into apish creatures that eventually changed into people (evolution). This totally leaves God out of the picture.

~or~

2. God created the universe and all life (including human beings) in six 24-hour days less than 6,500 years ago and there has been a world covering flood upon the Earth which rapidly and recently created the sedimentary layers and fossils which cover over 75% of the land surface on our planet.

When possible, demonstrations and experiments will be added to help you demonstrate which of the models for Earth history is the truth.

C. Two Models for Origins

Over the course of the last two hundred years, science has slowly changed from its original understanding as *"a systematic method of studying creation in order to understand the principles by which it operates,"* to *"the naturalistic understanding of the universe."* In other words, science, by definition, now cannot even consider the possibility of God's existence or interaction with creation anytime in the present, past, or future.

I vividly remember a conversation I had with a Ph.D. archaeologist over twenty years ago who had repeatedly insulted me in public letters to the editor in our local newspaper. At the time, I was writing a weekly column explaining the scientific evidence supporting creation, and this scientist accused me of undermining the Constitution of the United States and trying to drag our country back into the Dark Ages. I challenged him to a public debate on the issue and asked to meet him in his office to talk about the details. As we met, I explained that someone finding an arrowhead could choose to believe that random blows from other falling stones had created the arrowhead, but the symmetry and purpose of the object would make the truth obvious – it had to have been designed by intelligence. The complexity and interrelationship of the parts of even the simplest living cell makes this truth even more obvious. I further explained that I simply wanted students to hear all the facts so that they could determine the truth for themselves. This scientist got increasingly agitated as I talked, and by the time I finished, he stood up and literally shouted, *"Science is not searching for the truth (here he traced a capital "T" in the air), just the best natural explanation for things. If you want to search for truth - go home and do it, that search does not belong in a science classroom."*

This is not how modern science started or developed. Almost all of the founders of modern science, such as Isaac Newton, Joseph Maxwell, Blaise Pascal, Robert Boyle, Michael Faraday, Lord Kelvin, Louis Pasteur, Johannes Kepler, Gregor Mendel, Nicholas Copernicus, and a host of others believed in a recent creation by a Biblical God.[3] It was because the universe was created by a logical God that modern science could discover the principles by which it operates. Acknowledging this in no way prevents the discoveries of modern science.

Who Was Issac Newton?

Denying the existence of God guarantees that the past will be misinterpreted, often with tragic results. Based on the misplaced belief in human evolution, many scientific errors have slowed the progress of science. For instance, assuming DNA was filled with useless leftover "junk" programming stopped researchers from looking for functions for this programming. In the late 1800's, it was widely accepted that over 100 parts of the human body were useless leftover "vestigial" organs (i.e. tonsils, appendix, thyroid gland, etc. ...) Which needed to be removed whenever they caused a problem. This led to the unnecessary removal of these and other organs, resulting in thousands of needless deaths and illnesses.

Many scientists accept Biblical creation

The universe does operate by natural laws and it is extremely rare that God overrides these laws. On those few occasions when God has intervened in creation, He has made it abundantly clear in the Bible by stating so clearly and repeating it multiple times in both the Old and New Testaments. If God constantly performed miracles to change the way creation operated, we would have no confidence that things would happen tomorrow in the same way they happen today. Science would be impossible and unpredictable. But there have been at least three MAJOR interactions of God in the past, and He has made it emphatically clear that He, not natural laws of the universe, was responsible for these specific events.

+ + + + +

+ + + +

D. Why is Evolution so Widely Accepted?

You will notice that examples of evolution (with its required huge time spans) are the only option presented to students in the vast majority of textbooks and science curriculum's. Standardized testing requires that students answer science questions from an evolutionary perspective. Your textbooks are written by Ph.D. professors who have spent their life studying science in their area of expertise. A recent poll of biology professors revealed that 95% believe in the common descent of life (bacteria turning into people) as a <u>fact</u> of science. Essentially every geology department in the world teaches that the earth has been forming and changing for almost 5 billion years, and reject the possibility that there has been a worldwide flood upon the Earth. All of this ignores the creation interpretation of history (and the evidence which supports this), and directly contradicts the clear, emphatic statements found throughout the Bible. Many (if not most) teachers simply assume that since the majority of experts and textbooks teach about evolution (and enormous time-spans of Earth history), it must be true. After all, how could so many experts be wrong?

Ban Water

History shows that *"Consensus thinking, experts, and majorities"* in every field of human endeavor have been completely wrong in the past about science, morality, politics, and reality multiple times. Just in the last 200 years we have witnessed:

- Humans born with dark skin pigment were owned as property throughout Western civilization because of the color of their skin and their birth heritage. (Slavery)
- Today experts are encouraging confused men, women, and even children to undergo hormone treatment and surgery in an attempt to change men into women and women into men. (Sex Change Operations)
- The scientific consensus in the late 1900's was that the Earth was entering another ice age, we would be running out of oil by the beginning of the 21st century, overpopulation was going to cause worldwide famine, and an ozone hole and acid rain was going to destroy the environment. All of these scientific predictions were completely wrong. (Climate Change)
- Women being convinced to kill their babies as a perfectly acceptable alternative to giving birth. (Abortion)

There are many more examples of tragic and absurd beliefs of both scientific *"experts"* and *"the majority"* in present and past cultures. But one of the most prevalent beliefs of "the majority" is the idea that everything made itself and bacteria have turned into people (evolution). Think about what evolution is proposing as the reality for explaining our existence

- Order and design within the universe came from an explosion (the Big Bang)
- Rocks (i.e. chemicals dissolved in water to release their chemical constituents) came alive
- Information (the DNA code) wrote itself
- Random mistakes (mutations) cause things to improve
- Bacteria changed into apish creatures which changed into humans
- The intricate design of life is just an illusion and life had no designer

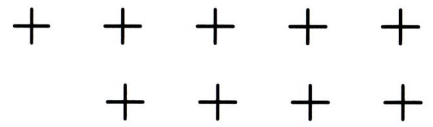

11

One of our jobs as Christians is to be a light of truth to help others examine the evidence in order to determine if these beliefs about evolution are true or not - rather than just blindly indoctrinating generation after generation with the beliefs of the "experts" who were likely themselves only shown an evolutionary viewpoint during their education. But the question remains, how can scientific experts promote and believe these obvious falsehoods? This did not happen overnight, but has been the result of a slow hundred year-long transformation within the scientific community. There is a vocal and active minority of scientists who look at the same rocks, biology, genetics, and laws of science and come to a completely different conclusion about our origin, but their voice is ridiculed, suppressed, and largely excluded from educational input.

There are four primary reasons why the majority of scientists and educators believe and promote only an evolutionary perspective:

1. They have been taught to view things from man's perspective rather than God's perspective. By defining science as only allowing the examination of natural causes, one of the two possibilities for our existence (creation) is not even considered. Thus, the only alternative to evolution (creation) becomes forgotten (and even ridiculed). When the truth is assumed to be wrong before even looking for an answer, believing in the wrong answer is guaranteed. The Bible makes the same observation:

> "The first one to plead his cause seems right, until his neighbor comes and examines him." – Proverbs 18:17 (NKJ)

2. They have been repeatedly told false information (meanwhile the truth is censored or ridiculed).[4,5] When generations of public educated children are given only evolutionary explanations for life, they form a framework of "truth" that comes to accept that belief as reality. They may not even realize that they need to even look for an option B. They literally become blind to any other possibility. As these students become the next generation of scientists and teachers, they teach that viewpoint even more strongly. Once a framework of what is the truth forms in the human mind, it become very difficult to view things in any other way. Even when shown that a belief could not be true, it is still accepted as true. The Bible calls this, "hardening of the heart." Here are a few pertinent observations from both science and the Bible:

> "Nothing is too absurd to be believed, if it is simply repeated often enough." – Dr. William James (father of modern Psychology)
>
> "The heart is deceitful above all things, and desperately wicked: who can know it?" – Jeremiah 17:9
>
> "All we like sheep have gone astray; we have turned everyone to his own way..." – Isaiah 53:6
>
> "The wrath of God is revealed [against] men who suppress the truth...that which may be known of God is obvious... being seen from the creation of the world...by what has been made...so they are without excuse...professing to be wise, they became fools." – Romans 1:18-22

Here are three wonderful examples of how the human mind can become literally blinded to the possibility of thinking in any other way:

Inverted Glasses
Show this video to your class at some point to help them understand how when their brain is taught to interpret things in only one way it becomes literally impossible to see something in any other way, even if what they currently believe is not the truth. Even our eyes can be trained to see things in a completely wrong way.
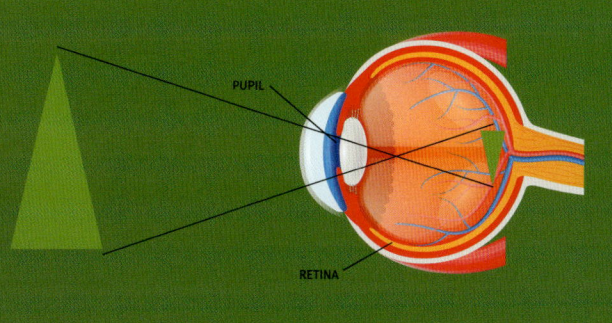

The Ames Window
Show this video to your class at some point to help them understand how when their brain is taught to interpret things in only one way it becomes literally impossible to see thing in any other way, even if what they currently believe is not the truth.

Backwards Bicycle
The backwards bicycle shows how once trained to do something in one way it takes enormous effort to change, even if you know how to change your thinking.

Student Demonstration - Optical Illusions

Show a large picture of the classic "old woman/young woman illusion" to your class. You can find this picture on google images listed under "old woman young woman illusion."

Ask your class:
- How many see an old woman?
- How many see a young girl?

Point out how to see each figure by highlighting and outlining specific features:
- Point to the "big nose and dark horizontal line as the old woman's lips" to see the drawing as an old woman.
- Point to the "tiny nose, chin and dark horizontal line as a necklace on her thin neck" to see the young girl

Now ask:
- By concentrating on specific features can you see both viewpoints?
- Can anyone see both viewpoints at the same time? (This is actually impossible)

Help them understand that when we are trained to see things a certain way, seeing them any other way becomes increasingly difficult and it is impossible to truly believe two contradicting beliefs at the same time. You will ultimately allow your life to be guided by the one you truly believe.

3. There is enormous pressure to conform. All people have a tendency to be influenced by the opinions of others. No one is immune to this. This is why companies spend billions advertising products and polls are widely publicized showing what is most popular. Scientists are also influenced by consensus thinking. Science which deals with the past (geology, paleontology, cosmology, biological origins) is not directly testable and reproducible. Therefore, it is highly speculative. Yet "speculative science" is frequently presented as "testable science," i.e. the laws of physics or chemistry. Even more tragic, as molecules-to-man evolution has become accepted as a scientific fact, any teacher or professor who dares to teach students an alternative will often be fired or removed from their job.[6,7] Thus, even if they have doubts about what they are required to teach, they often remain silent.

4. Researchers must be paid. An enormous amount of money is spent to promote evolution. Tiny amounts are spent to promote creation. The only research that gets serious funding is for evolution. Furthermore, any researcher who promotes a creation viewpoint, or even openly questions evolution, will not get their papers published and risks their academic career and future earning ability.

Bias when using Science Observation

KEY POINTS TO REMEMBER: + + + + + + +

1. There are only two possible explanations for our existence – evolution or creation.
2. Most of the founders of modern science believed in creation, and this belief did not slow scientific advancement.
3. Evolution can mean different things. When it means organisms are programmed to change and adapt, it is true. When it means one animal type can change into a completely different body structure, molecules-to-man, it is faith, not science, and is not observed in science, therefore it is not true.
4. Science has been redefined to not allow the consideration of creation as a possibility.
5. Majority opinion does not determine truth, and scientific consensus has often been wrong.
6. Textbooks ignore the evidence for creation because:
    A) God is assumed to have never been involved.
    B) Scientist have been trained to think in only one possible way (evolution is a fact).
    C) There is enormous peer pressure to conform.
    D) There is relatively little funding for creation research and promoting creation can get a researcher or teacher fired.

SECTION II - PROBLEMS WITH CHEMICAL EVOLUTION

A. Could Chemicals Come Alive?

For life to evolve from bacteria to people, scientists must first explain where bacteria came from. This is commonly referred to as "chemical evolution." Atheists confidently promote the chemical evolution scenario proposed in textbooks as a fact of history and assume it explains how life could have formed on earth. Ignored is the reality that each step which has been proposed for the chemical evolution of life is demonstrably impossible. Since God is left out, and life does exist, it is simply assumed that life popped into existence from nonliving chemicals.

The reality that life never creates itself has been confirmed by countless experiments.[8] Even the most famous origin of life experiment, the Urey/Miller experiment (discussed in the next section), actually shows that life could never make itself because the correct chemicals needed for life cannot randomly produce themselves. Life is too complex to ever make itself – life exists because God made it. An increasing majority of scientists simply refuse to acknowledge the obvious, and textbooks reflect their belief rather than teaching the truth.

When approaching this subject with students, keep the bigger perspective in mind - using all of our technology and knowledge of exactly how a cell operates...and using the biochemicals needed for life to exist (which are only made inside of already living cells)...every experiment ever done confirms this fact of science – chemicals NEVER come alive by themselves. Even if we did succeed in producing life in a laboratory, it would only prove that intelligence is required, not that life could make itself. And we are not even remotely close to making this happen. We can modify pre-existing life, but never create life.

Life Isn't From a Lab

The Miracle of Life

B. Problems With Origin of Life Experiments

Let's take a closer look at typical textbook statements about the origin of life. Your students should not only be aware of what the textbook writers are trained to believe, but what is being left out of the discussion. Here are some very common statements used to teach students that life made itself.

> **Misleading textbook statements about the origin of life[9]:**
>
> 1. The earth formed 4.6 billion years ago, and life first appeared 3.9 billion years ago.
> 2. Chemicals present on the early earth assembled themselves to form cells.
> 3. It is implied that wherever water is available life can form.
> 4. The early earth atmosphere had no oxygen, this appeared as plants produced oxygen.
> 5. Miller's experiment shows how the chemicals needed for life could form themselves.
> 6. Amino acids formed into proteins in shallow pools of water.
> 7. RNA formed the template to create reproducing life forms.
> 8. Proteins and RNA naturally form into spherical "protocells."
> 9. The chemicals of life may also have come to earth from other planets.

1. The earth formed 4.6 billion years ago, and life first appeared 3.9 billion years ago.
It is an assumption of cosmic evolutionary processes that leads to the belief that the earth is 4.6 billion years old. There are enormous problems and inconsistencies with radiometric dating methods which yielded this number, and these will be discussed in detail in a later section. For now, informed people know that all methods for dating the distant past are based on unprovable assumptions and there are many more methods indicating a recent creation that contradict this widely publicized age for the earth.

2. Chemicals present on the early earth assembled themselves to form cells.
The chemicals needed for life are very specific and only occur inside of living cells. No scientist has ever found biologically useful proteins, enzymes, lipids, carbohydrates, polypeptides, or DNA fragments forming outside of a living cell. Chemicals which do form naturally are toxic to life and are not useful for forming life.

Whenever formation of life experiments are done in laboratories, the scientists start with purified chemicals which were formed by life. Thus, they are "cheating" in their experimentation from an invalid starting point. In spite of this "advantage," every experiment ever done by every researcher in every laboratory, after hundreds of years, thousands of attempts and costing millions of dollars, has ALWAYS had the same result: chemicals never come alive by themselves. <u>If they did make life, then why aren't scientists using this technique to routinely develop solutions to our myriad of problems?</u> Science and experimentation clearly reveal the truth – life (and the chemicals required) had to have been organized by an incredibly intelligent designer.

It is widely taught that bacteria were the earliest forms of life on Earth and therefore were the simplest. This link shows the complexity of a bacterium called MO-1 <u>which has been found in the lowest layers of rock containing fossilized life on Earth.</u> This bacterium contains 7 green gears (driven by miniature motors) connected by 21 smaller gears driving hair-like cilia to create miniature propellers! What drives the gears? Miniature motors made of specifically shaped protein parts so small that 1000's of these motors could fit across the width of a human hair. Hardly a "simple" form of life!

The MO-1 Bacterium

3. Wherever water is available life could form.

This is implied by textbooks and widely promoted by the media. Whenever water is found on planets in our solar system or proposed to exist on planets orbiting other stars, it is stated that "life could form there." This is analogous to telling students that because iron ore is found in a rock layer, given enough time, an automobile could appear all by itself. We all have water in our kitchen sinks, but new forms of life are not in the process of forming. These types of statements are religious beliefs, not science. They are actually refuted by scientific observation. Also, life has never been seen to form in any body of water all by itself - whether in oceans, lakes, ponds, swamps, laboratories, or kitchen sinks. It just NEVER happens!

Frog Blender

4. The earth's early atmosphere had no oxygen.

No scientist has a time machine. This statement is an assumption based on the need to explain the origin of the most abundant type of molecule within a cell (amino acids) by natural processes. These chemicals cannot form in the presence of oxygen, so it is ASSUMED there was no oxygen in the early earth atmosphere. But what do observations show? As far down as we go into the earth, the rock layers show oxidized chemicals. In other words, as these layers were forming, there was free oxygen in the atmosphere for them to react with. Oxygen has always been present in the earth's atmosphere.[10]

5. Miller's experiment shows how the chemicals of life formed all by chance.

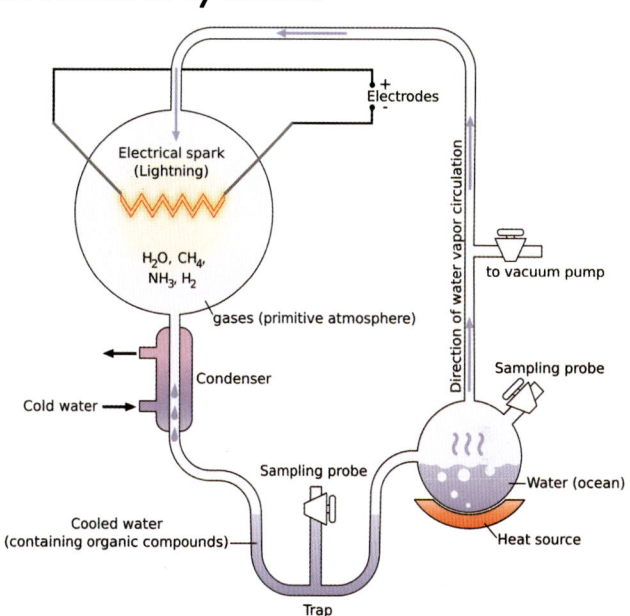

GYassineMrabetTalk This W3C-unspecified vector image was created with Inkscape . W3C validity not checked. (https://commons.wikimedia.org/wiki/File:Miller-Urey_experiment-en.svg), "Miller-Urey experiment-en", https://creativecommons.org/licenses/by-sa/3.0/legalcode

Essentially every biology textbook in the world promotes the Miller/Urey experiment as proof that life could make itself. In 1952, a researcher circulated methane, ammonia, hydrogen, and water vapor past a high energy source (simulating lightning) and produced a few types of amino acids. Amino acids are called the "building blocks of life" because they link up to form proteins. Proteins are the most common molecules in a cell and are specifically shaped, like the designed parts of a machine, to perform a myriad of specialized functions within a cell. It is the specific order (or arrangement) of the amino acids which allow the formation of the exact shape needed for each protein to function. All forms of life result from the precise arrangement of 20 specific types of amino acids. This arrangement forms the three-dimensional shape of the needed protein. The simplest bacterium requires 1,000 different and specific protein structures within its cell.

+ + + + +

+ + + +

19

The human body requires an estimated 100,000 specific and different proteins. Miller's experiment made only 5 of the 20 required amino acids, yet students are told that this is how life could have formed here on earth. This is what textbooks leave out and students are NOT told:

Magic can of Evolution

- All 20 amino acids are needed for life but only 5 were formed by this experiment.
- All Amino Acids are either left-handed or right-handed. Living organisms only use left-handed molecules in their cells. Therefore, all protein chains must be 100% left-handed for life to function. The Miller/Urey experiment made both, which are impossible to naturally separate. Thus, the chemicals made in this experiment are useless and detrimental for forming life.
- Other chemicals were formed, like formaldehydes, that would have stopped the amino acids from linking up into protein chains.
- Ultraviolet light from the sun would have destroyed any useful proteins long before any living cell could form.
- The idea that life can 'spontaneously generate' from non-living chemicals was disproved by Louis Pasteur more than 100 years ago. The law of biogenesis is an observable and testable law of science that states, "Life comes only from previously existing life," i.e. it could never make itself. The Urey/Miller experiment is just a modern version of the discredited spontaneous generation belief. Biogenesis has been scientifically established to be a LAW of science - no exceptions have ever been observed

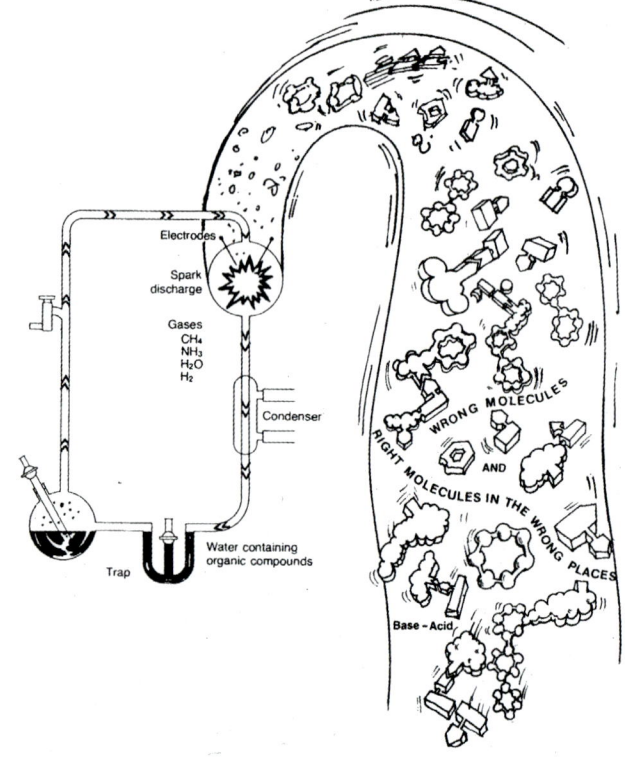

http://www.verhoevenmarc.be/PDF/Abiogenese.pdf

What Miller's experiment actually shows is that life could not possibly have formed in this way.[11,12] Yet it is still promoted in textbooks 70 years later and the enormous problems with this as an explanation for life are hidden from students. Why? Because nothing better has been found and it is assumed that God did not create life - so students are *led* to believe that something like this must have happened and the problems with this belief are not mentioned.

Student Demonstration - Can an Explosion Create Life?

Stories, suspense, and surprise will be remembered far longer than facts. Use this demonstration to show the impossibility of Miller's experiment creating life:

+ + + + + + + +

The Magic Can of Evolution Demonstration

> Preparation:
> Find a large coffee can with a plastic snap-on lid and cut a 3/8" (10 mm) hole in the lid.
> Find 2 identical pens which can easily be disassembled into 5-8 parts.
> Purchase a package of firecrackers and butane lighter.
> Loosely tape one of the pens into the side of the can and put the rest of the items into the bottom of the can.

- Tell the students that the simplest single-celled organism, such as bacteria, has thousands of specifically designed parts. Explain that life is far more complex than a pen, but *"Let's see if a pen could create itself."*

- Dump everything out of the can (don't let the students see there is another pen taped inside) and explain that the pen is made of many parts designed to become a pen. Not even the parts could create themselves by accident – just like the parts of life are designed to become a living cell. These complex biochemicals are not found outside of living cells.

- But even if you have the parts, could a pen form? Take apart the pen and put the parts back in the can. Shake the can. Spin the can. Throw the can in the air. Ask the students, "Do you think adding lots of energy will create a working pen?"

- Explain that Miller's experiment used explosive energy, so you are going to try that. Light the firecracker and drop it in the can. The lid will blow off and smoke will come out. Then dramatically act surprised and pull out the whole taped pen while leaving the disassembled parts in the can.

- When the laughter stops, shake the can so the students can hear the disassembled pen inside. Say, *"I tried to fool you, but you knew a pen cannot make itself. I have two pens."* A pen has just 6 parts while the simplest cell has more than a thousand parts. If a pen with six parts cannot assemble itself how could a simple cell? The simplest cell has more than a thousand parts – it is billions of times more ridiculous to believe it could make itself!

Practice this before doing it in the classroom. I have done this demonstration with over 100,000 students and even ten years later people still remember the lesson learned.

+ + + + + + + + + + + + + + + +
+ + + + + + + + + + + + + + + +

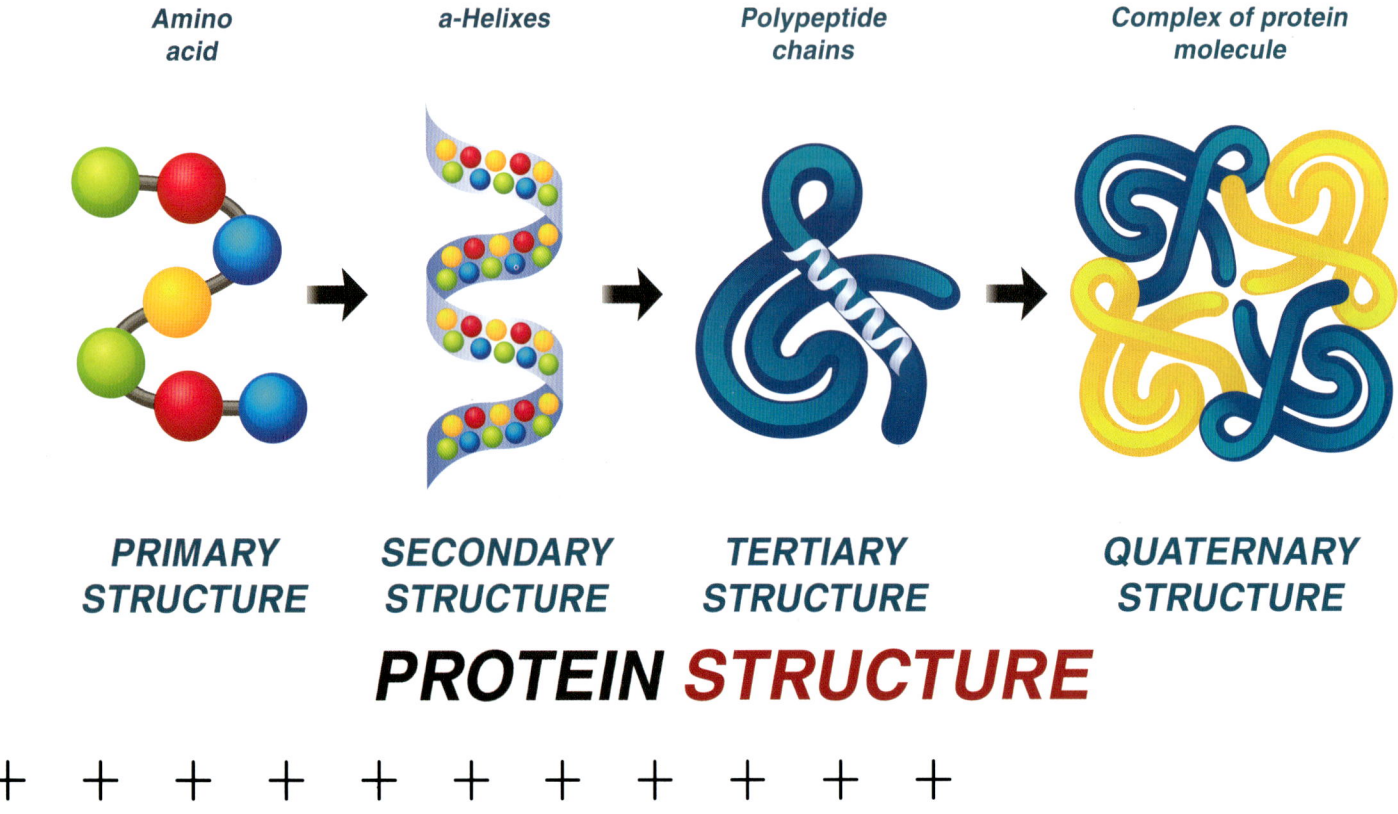

PROTEIN STRUCTURE

+ + + + + + + + + + + +

6. Amino acids formed into proteins in shallow pools of water

Amino acids can randomly link up to form proteins, but it is impossible that these randomly produced proteins would be useful for life. The amino acids must line up in a very specific order to form a biologically useful protein. Otherwise, the protein does not have the correct shape (like the wrong part in a machine) and a cell would never function.

In many ways a living cell is like a complex machine made from many specifically designed parts. The most common chemical in any living cell is a protein. Each protein in a cell is made by lining up 20 smaller molecules, called amino acids, in a specific order so that the protein will assume an exact 3-dimensional shape – not unlike the parts of any machine or the exact shape of a key needed to work in a specific lock. A typical cell has at least 1,000 different specifically designed proteins (parts). **To randomly form the correct sequence to make even one correct protein (machine part) will never, ever happen.**

The average protein is 400 amino acids long. The probability of one average length, 400 amino acids long protein, randomly lining up in the correct order (even if all 20 amino acids were available) is once every 10^{250} tries (once in every 10 with 250 zeros times). It is estimated that there are only 10^{80} electrons in the entire universe. If the universe were 14 billion years old (10^{18} seconds), and every particle in the universe interacts with another a billion times per second (10^{12}), there still would only have been 10^{100} interactions (or chances) to make one of the correct proteins needed for life. Making even one of the one thousand proteins needed for even a single bacterium to form by natural evolutionary processes is the very definition of impossible! Statistical science confirms that the specific proteins needed for life would never happen via evolutionary processes.[13]

Student Demonstration - Can Useful Information Create Itself?

For useful molecules of life (proteins) to form, amino acids must be arranged into the correct order so the right shape of molecule can form. In a similar way, for useful information (sentences) to form, letters must be placed in the correct order and a meaning must be assigned to the words. Do an experiment to see if this could happen.

- In front of the classroom, show the students a large piece of paper, preferably cardboard, with your school's name printed on it in large block letters.
- Now cut it into pieces. Each piece should have only one letter on it.
- Select one student and place all the letters in that student's hands.
- Have the student drop all of the letters at once from 3 feet up. **Did the letters spell out your school's name? Did the letters line up to form any other words?**

Explain to the students that maybe falling from only 3 feet did not give random chance enough <u>time</u> for the letters to line up correctly.

- Select another student and have this person stand on a chair and drop the letters. **Did the letters spell out your school's name? Did the letters line up to form any other words?**

Maybe falling from only 5 feet still did not give random chance enough <u>time</u> for the letters to line up correctly.

- Have a third student climb onto a table and throw the letters up toward the ceiling. **Did the letters spell out your school's name? Did the letters line up to form any other words?**

> Ask the students these questions:
>
> If we put the letters in an airplane and dropped them from a great height so that they had lots of time to form correctly as they fell, would they spell out your school's name?
>
> If we dropped the letters over and over again for a whole day, would they spell the name of the school? How about for a whole year? A million years?
>
> Even though we gave more time for the letters to line up, they did not. Random chance over time does not create useful or intelligent information. Tell them to remember this experiment when they are told that random chance (mutations) created the useful or intelligent information needed to make the molecules of life.

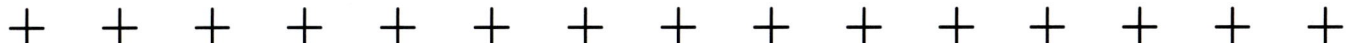

7. RNA formed the template to create reproducing life forms

DNA is the most compact information containing system in the universe. Scientists are just beginning to be able to read it. Delving into the DNA's information system can be mind-blowing! Let's use an analogy to give us a small insight into the DNA's sophistication. Imagine finding an electronic tablet in a deserted alien spaceship. On this tablet is an entire library. The linguists begin deciphering. First they discovered that the sentences could be read right to left AND had a difference meaning when read left to right. Then it was discovered that some of the instruction manual could be read in a different language. Like reading it in English to get the first ½ of the story and then starting from the beginning and reading it in French to get the second ½ of the story.

In addition, one of the languages consisted of only reading three letters. If you started at different positions and read, the meaning was changed each time. Needing a quiet place to work on deciphering the tablet, one of the linguists took the tablet into the kitchen. When she opened the document, she noticed some of the letters had been "grayed out." When she read only the black letters, another message was found. When the tablet was taken into the navigation office, other letters were "grayed out" and a different message appeared. Just like in this example, our DNA has an astonishing level of "data compression!" Scientists have only scratched the surface of deciphering DNA.

DNA requires specific proteins to do all this. How could this information system come together? Textbooks with an evolutionary viewpoint propose that RNA formed first and slowly turned into DNA. What is RNA? It is a copying, or messenger molecule that lines up alongside DNA, copying a form of the information onto the RNA molecule, and uses that information to make the proteins needed to construct components of a living cell.

There is absolutely no evidence, despite decades of experimental attempts, that useful RNA segments could produce themselves.[14] Furthermore, just like with fragile protein molecules, we know these RNA fragments would disappear rapidly after being produced - making it astronomically unlikely any could form outside of an already living cell. Finally, every attempt in laboratories to make living cells, even when starting with RNA and proteins from already living organisms, utterly fails.

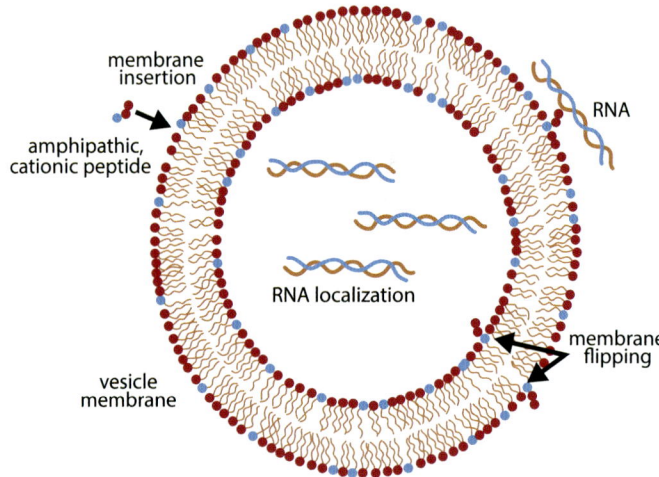

8. Proteins and RNA naturally form into spherical "protocells"

Many biology textbooks report a widely publicized experiment that shows that small protein chains, called peptides, when mixed in a water solution with RNA fragments, will forms small circular structures with the RNA encapsulated inside. Furthermore, these small structures can be shown to split and divide into multiple spheres. Students are led to believe this demonstrates an early form of cell formation and division.

These polypeptides have both a polar and non-polar end. Water is a polar molecule. The polypeptides naturally separate exactly the same way oil and water separates, but when soap is added to water, it surrounds and dissolves the oil. This forms microscopic spherical structures. This happens naturally with polypeptides and the RNA (which is attracted to the non-polar side of the polypeptide molecule) gets trapped inside of the "oil blob." Movement can cause these blobs to break into smaller blobs. This is not even remotely similar to a living cell and it is deceptive to call them "protocells." The fact that oily spheres can form in a water solution is promoted as an explanation for how life could form and is illustrative of just how little evidence there is to support chemical evolution.[15]

9. The chemicals of life may have come to earth from other planets.
The idea that life could have evolved from non-living chemicals is called abiogenesis. Having never been proven, a number of evolutionists have come up with another idea. Life came from outer space, called "panspermia." To test this idea, scientists in France attached rocks with a hardy bacterium onto a Russian spacecraft returning to earth. This was a simulation of a meteorite carrying life to our planet. The result: the bacterium was burned to a crisp. Upon entry into earth's atmosphere, the organism was burned black, becoming totally carbonized.[16]

Point out to students that the belief in aliens planting life here on earth, or life coming to Earth from outer space is faith, not science. Experiments to test this idea show it could not be true. There is absolutely no evidence to support this belief and it still does not overcome the problems with believing in evolution. If chemical evolution could not happen here on a planet perfectly designed and created to support life, it is even less likely to occur elsewhere.[16a] Yet the entertaining idea of aliens is everywhere in the movies, and as the father of modern psychology, Dr. William James stated, *"Nothing is too absurd to be believed, if it is simply repeated often enough."*

Brilliant Ancient Mankind and Alien Life

KEY POINTS TO REMEMBER: + + + + + + +

1. Every experiment to make life from chemicals shows that this never could happen.
2. Experiments to make chemicals come alive starts with chemicals from living organisms. This "cheating" still does not result in life forming.
3. Miller's experiment actually shows life could never form by accident.
   A) Not all of the needed amino acids were formed.
   B) The wrong kind of amino acids were formed.
   C) Other chemicals formed which would have prevented protein formation, i.e. formaldehyde.
4. Proteins must be in a specific order. This is impossible by random arrangement.
5. "Protocells" are just blobs of oil and are not remotely similar to the complexity of a cell.
6. RNA copies information from DNA and the source of any useful information, language, or code is ALWAYS intelligence, not random changes.
7. Finding water does not explain how life could form.
8. There is no evidence that alien life exists or that it could make itself.

SECTION III - PROBLEMS WITH BIOLOGICAL EVOLUTION

A. The Classifying of Life

There are many examples in textbooks "seeming" to prove that biological evolution is a fact of science. In this section we will discuss this "evidence." This enables teachers to separate fact from speculation. Remember that life is designed to adapt and diversify. In your textbooks, you will find many examples of life's diversity and ability to adapt. This is used to promote biological evolution as a fact of science. As students are presented with this evidence for evolution, continually frame the evidence in a perspective of the two larger questions:

Where did this ability to adapt and diversify come from?

Could this process change one organism into a completely different organism, i.e. bacteria into people?

Biological evolution is such a broad subject that our discussion of it will be broken into specific areas with the common textbook evidence supposedly supporting evolution discussed within each broad category. Your students should not only be aware of what the textbook writers are trained to believe, but what is being left out of the discussion. If you have time, show this short video on the wonder of life. Use this as an example of how even instincts were created by God. Why else should such creativity and beauty exist?

a) The Linnaeus system of classification

Carl Linnaeus (1707–1778) was a Swedish scientist who brought into general use the system of classifying all living organisms with two Latin words. He was able to name thousands of plants and animals and became the "Father of Taxonomy." We still use this system today.

Textbooks state that the classification of plant and animals reveal a progression of evolution from simple life to complex life and ultimately to apelike creatures which evolved into modern humans. The Linnaeus universal classification system, used for categorizing all biological life, was developed by a brilliant Christian _who believed in a recent creation of very different forms of life_. Obviously, the system of classifying life which he developed does not require the belief in evolution. In actuality, a belief in evolution would have hindered development, as evolution implies a blurred continuum of life rather than distinctly different created body structures.[17]

Linnaeus recognized that there is a lot of variation within major groups of animals, but rejected the belief that one organism could slowly turn into a completely different "_critter._" His entire taxonomy system was based on his belief "_that God could be approached through the study of Nature,_" and he felt it was his Christian obligation to learn about God by studying "_...the wonders of the created universe._"[8, 18a]

The Japanese Puffer Fish

KEY POINTS TO REMEMBER:

1. Variation and adaptation are programmed into a given type of creature.
3. Variation within a given kind of creature is limited to the information already contained within its DNA code.
4. The system of classifying all forms of life was developed by a creationist and has nothing to do with the evolution of life.

b) Do evolutionary trees prove evolutionary relationships?

When Darwin proposed the theory of evolution back in 1859, it was believed a single-celled organism was a simple tiny blob filled with chemical goo and that life could slowly transform from one type into another. Darwin proposed a systematic method by which this transformation could take place (natural selection) and put all forms of life into an increasingly branching tree starting with the simple single-celled blob of goo and ending with very different, and much more complex creatures at the tips of each branch. But in Darwin's proposal, all forms of life were related to some previous form, right back to that original single-celled organism, and absolutely no interaction by God was needed to explain the development of any form of life on Earth.

This tree of life is still shown as a fact of science to students throughout the world and the visual nature of the arrangement seems very convincing. However, it is extremely deceptive because what is left out. For instance, in the version of the tree illustration, notice the development of needed changes marked by white spaces (unknown spots) in the tree's main trunk:

- *Life can't form:* DNA, nuclei...and many more unexplained advances must occur.
- *Animals can't evolve:* organs, nervous and vascular systems, digits, hair, placenta...and many more unexplained advances must occur.
- *Plants can't evolve:* Chloroplasts, photosynthesis, seeds, and other major structures are needed. The individual steps needed for the organism to advance are simply missing.

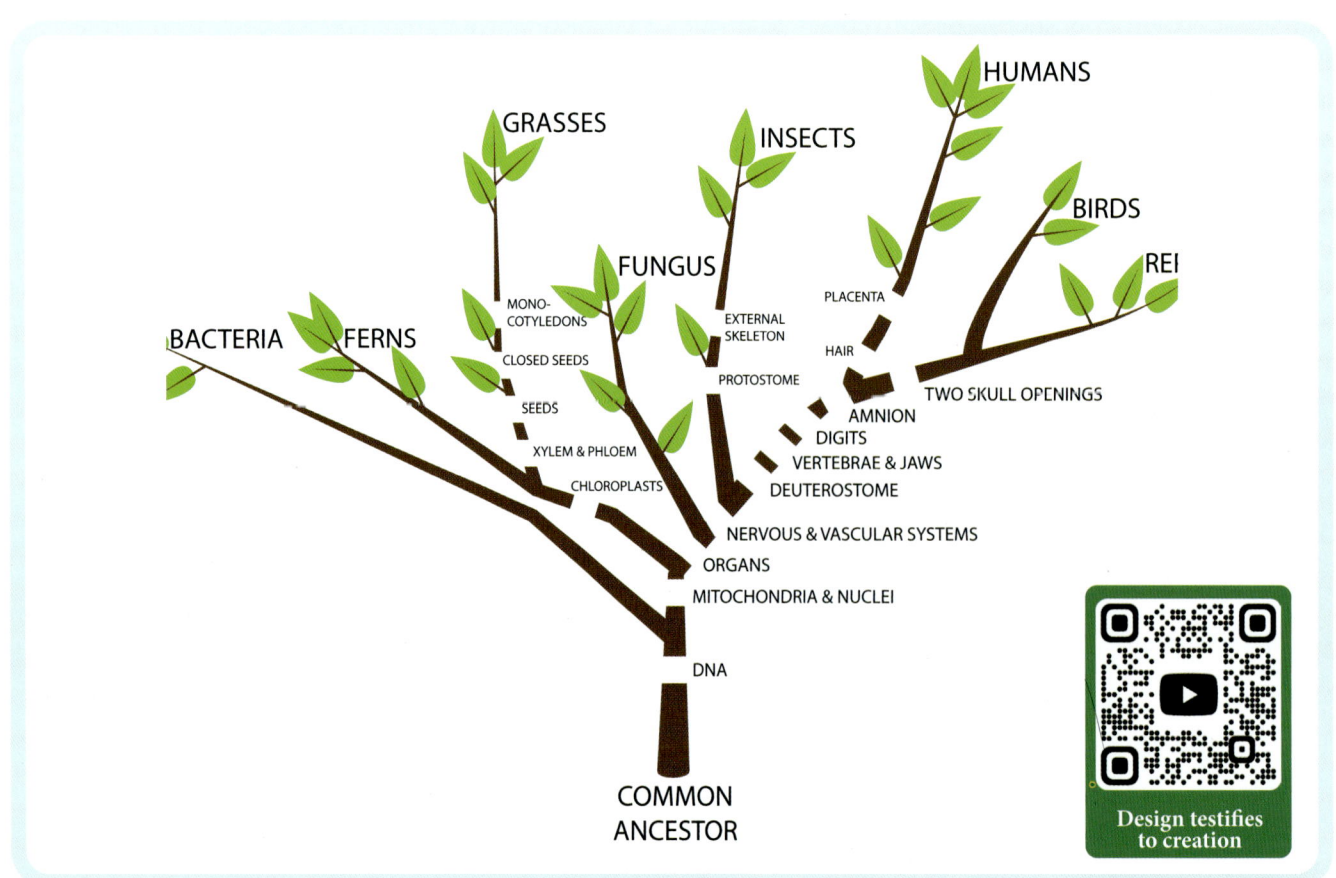

NONE OF THESE MAJOR, COMPLEX CHANGES HAVE EVER BEEN EXPLAINED BY EVOLUTION.

Student Demonstration - Photosynthesis

For instance, ask your students:
- What purpose would a partially formed eye. lung, heart, stomach or liver serve? How could a transitional animal survive which had only evolved part of a heart?
- Of what function is a brain without a nervous system?
- How could blood cells carry oxygen to the body if veins/arteries had not yet evolved?
- Why would an animal which functions perfectly without arms or fingers evolve the thousands of structural changes needed to form fully functional arms, legs, or digits?
- Photosynthesis (using sunlight to turn carbon dioxide, oxygen, and water into the chemicals needed for the cells within living plants) involves dozens of chemical reactions, each step depending on all the other steps being in place. How did all these reactions develop simultaneously? Any one is useless unless all are present.

More Complex Than you Think

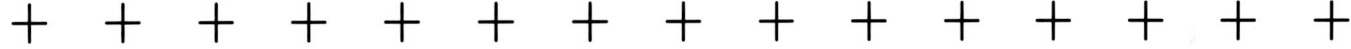

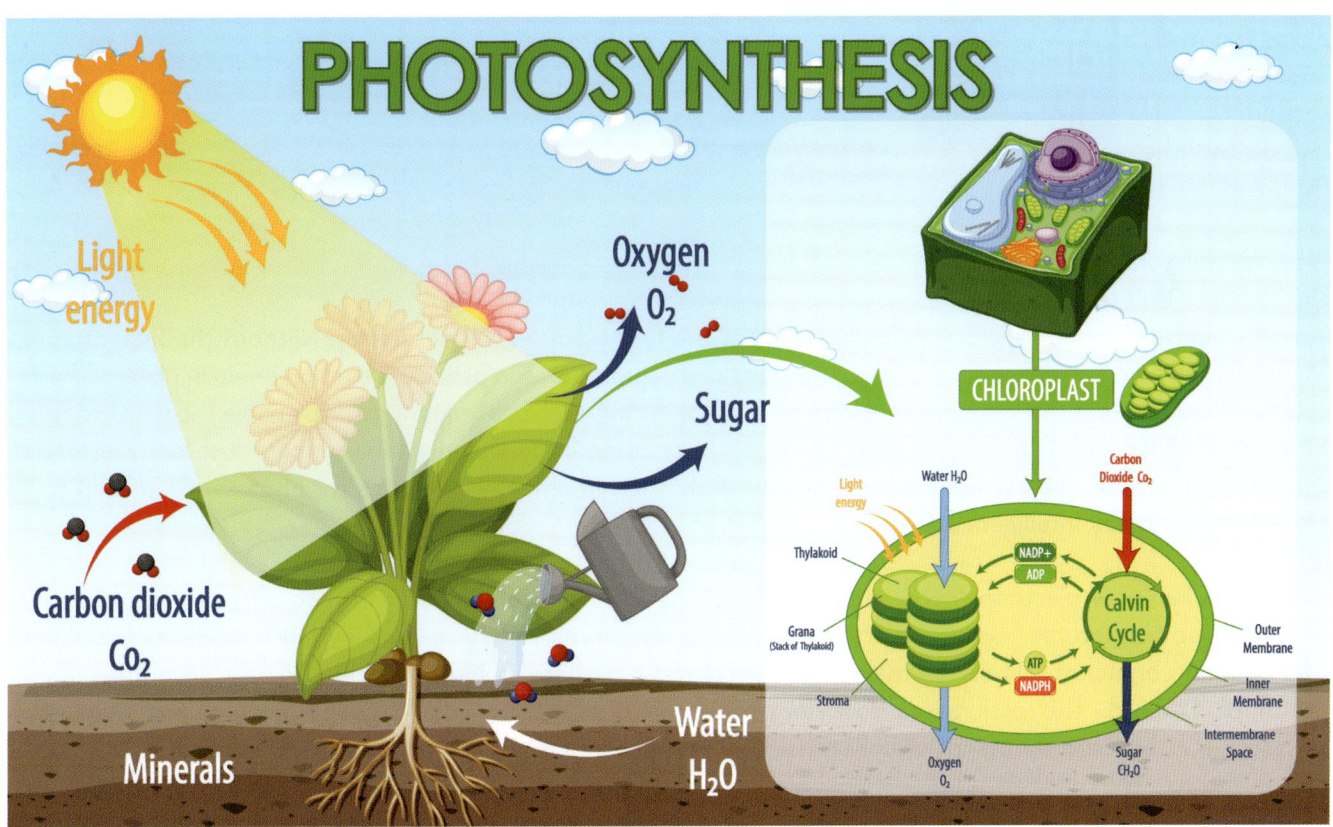

You can find photosynthesis diagrams on the Internet or draw simplified versions on the board. A key to plant evolution would be its ability to use sunlight to change water and CO_2 into sugars. Line up a row of four tall blocks representing dominoes (tall blocks of wood would work). Explain that each block represents one of the complex reactions used to make the sugars and other chemicals needed by plant cells. Knock down the first block and it will knock down the second, followed by the third and fourth.

Too Complex for Evolution

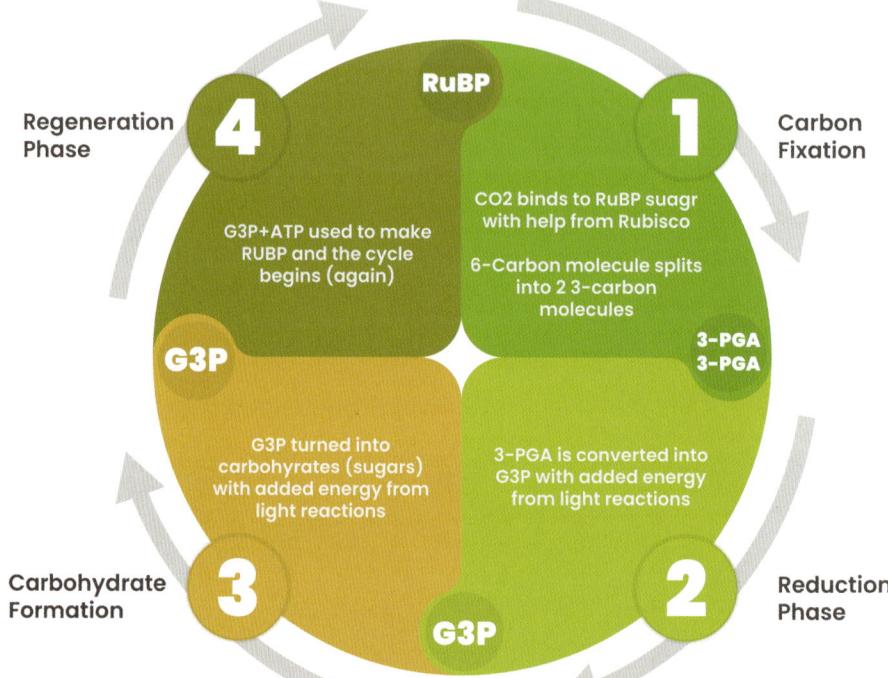

Now remove one of the blocks in the middle. Will the last block fall - producing the needed sugar? Obviously not. Yet the first reaction will not happen without the final product, sugar, already being present. Each of the four reactions depends on the previous reaction being in place. Chemicals previously made from the sugars need to be in place during each subsequent step to enable the next reaction. How could all the dominoes and needed chemicals have come into existence all at once? Just like you placing the blocks in exactly the correct position on your desk, the sequence of reactions for photosynthesis had to be placed in position by intelligence.

At essentially every place on the evolutionary tree where a trunk breaks off into two different branches, i.e. different body structures "evolve," there should be a question mark. These "missing link" creatures (not found wherever new branches split off the main trunk) are fantasy, are hard to even imagine, and have never been found in the fossil record. The fossil record does not contain just one "missing link" between mankind and some apish ancestor, but there are thousands of links missing. **They are "missing" because they never existed!**

Sun Tracking Sunflowers

Honey Bee Brains

Leaf Hopper gears

Slug Slime Band-aids

Sea Sapphire Invisibility

Starfish Stomach

Spider Abilities

Student Activity - The Tree of Life

For a student group activity, show or draw on the board a simple version of the tree of life (or refer to one in the textbook, pg. 32). Then ask your students to draw some unknown ancestor (parent) to one of these supposed evolution transitions. Show them an example of the idea that a cow evolved into a whale.

- A fern into a pine tree
- Grass into a palm tree
- A sponge into a jellyfish
- A beetle into a lobster
- A starfish into a fish
- A cow into a whale
- A crocodile into a bird

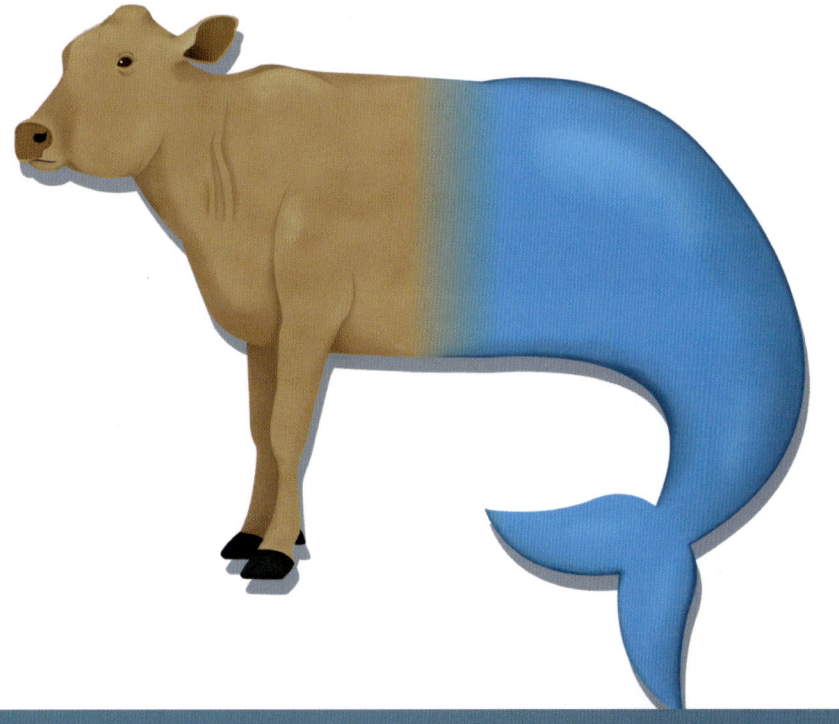

Was what they drew fantasy or scientific fact?

What we actually find, in both the natural world and the fossil record, are distinctly different body structures of specific animals with no clear transitions between these distinctly different groups. There can be lots of variation within a "kind" of creature (look at the wide variation of dogs), but dogs have always been dogs and we never find a catdog (part dog/part cat.) Life is never represented by a single tree but rather by an orchard – lots of distinct kinds suddenly appearing with the ability to adapt and form many branches and variations. Examples of this ability to vary is presented as proof of evolution but this ability to adapt in order to survive in different environmental niches DOES NOT explain where the distinct and different trees came from to begin with. Below is the evolutionary concept of the tree of life compared to Biblical creation's orchard of different created kinds. Each kind can have lots of variation but maintains the same basic body structure. This is actually what science and the fossil record shows.

Imaginary Evolutionary Tree

KEY POINTS TO REMEMBER:

1. Lining up similar animals does not prove one turned into another.
2. There is a systematic pattern of missing links between very different body structures in the fossil record.
3. Part of a new creature, organ, or cell would be a detriment to life, not something that allowed development of a new creature.

Imaginary Evolution Tree

Factual Evidence from Fossil Record

C. Do Similar (Homologous Structures) Show Evolutionary Relationships?

A classic illustration of the "similarity" between the bones in the arm of a human, bat, and whale is shown in almost every biology book in order to convince students that because these bones have similarity in shape, it proves that they all had a common ancestor.

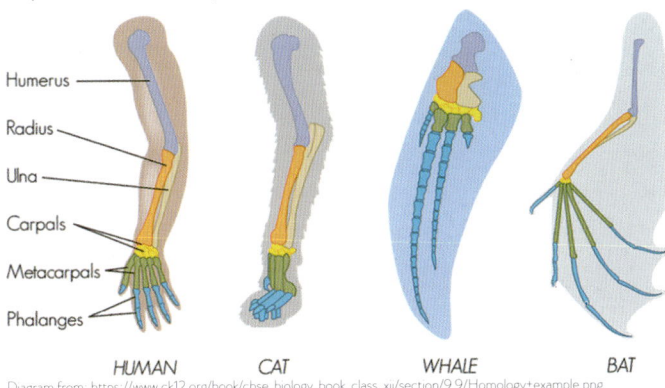

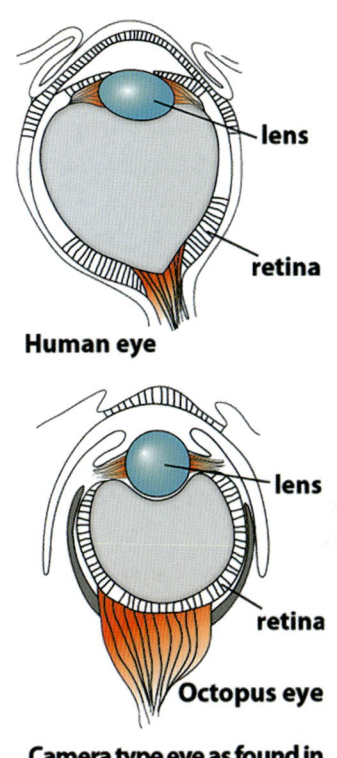

Human eye

Octopus eye

Camera type eye as found in humans and octopuses.

There is another view - same design, same Designer. That Designer designed for the same function, for example four legs for cats, horses and other land mammals. Same design for the same function. Similarity does not prove that different animals had a common ancestor, just the same designer. These distinctly different creatures have similar bones because it is an elegant and useful design and the same designer used modified versions of this design for different functions in different creatures.[20,21]

- Reaching for humans
- Walking for cats, horses, or other land mammals
- Swimming for whales
- Flying for bats

Even more revealing is that evolution believers pick and choose which homologous features to accept as "evidence" for evolution. For instance:

The eye of an octopus is essentially identical to a human eye - yet no-one claims we are the close ancestor to an octopus. Since this idea does not fit the evolutionary belief, it is not shown in textbooks as an example of homologous features supporting evolution.

- A platypus has a bill/webbed feet like a duck, lays leathery eggs like a lizard, has a poison gland like a snake, and hair/wide tail like a beaver. Yet no-one claims this animal is the direct relative of a duck, lizard, snake, or beaver.

Photo credit: Yarra Ranges Tourism

If structures were homologous, you would think they would develop the same way; but that is not always the case. For example, how do the digits develop on a human vs. a frog? Fingers on a human develop first as a paddle then cell death removes the material between the fingers. Compare this with a frog's digit development where each digit grows out from buds. This is a totally different mechanism requiring enormously different DNA programming. If one system worked fine for the frog, why would a completely different system need to evolve? They are both vertebra with hands having five digits. Why aren't these two similar if evolution is true and they had a common fish ancestor?

Painter's routinely copy other painter's styles. Scientists build on the work of other scientists, and architect's use the same principles to design buildings and bridges. Their resulting work has many similarities, but it is always a result of the intelligence and creativity of the designer. Homologous features are evidence for a <u>common designer of life</u>, not common ancestors for all of life

KEY POINTS TO REMEMBER:

1. Similar features do not prove ancestral relationship.
2. Similar features which do not fit the evolution story are ignored. i.e. Octopus & human eye.
3. Complex useful designs require a designer.
4. Similar designs are evidence for a common designer, not random change. God designed the best parts and used them repeatedly in his various biological designs.
5. If homology from a common ancestor were true, then the embryonic development should be the same for each animal, but it is not. i.e. the human and frog digit.

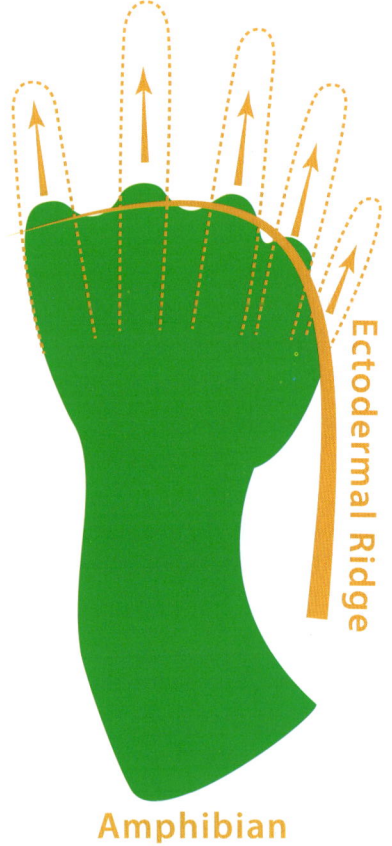

Human   **Amphibian**

Diagram From: https://phys.org/news/2015-06-feathered-dinosaurs-complex-thought.htmlhttpsphys.orgnews2015-06-feathered-dinosaurs-complex-thought.html.jpg

D. Have Dinosaurs Turned into Birds?

It is routinely taught that birds exist because dinosaurs slowly evolved the ability to fly over time. Evolutionists believe that dinosaur scales slowly turned into feathers and their front legs turned into wings. It is claimed that some fossils of dinosaurs have been found with feather-like structures, and the bones of upright walking dinosaurs (such as a T-Rex) are very similar to the bones of birds. Is this a fact of science or just another belief driven by an assumption that evolution must be true?

One of the classic "proofs" of evolution, which has been in textbooks for over one hundred years, is the discovery of a fossil bird called Archaeopteryx (meaning, "ancient wing.") According to most textbooks, this animal is the oldest known fossil bird, of the late Jurassic period. It had feathers, wings, and hollow bones like a bird, but teeth, a bony tail, and legs like a small dinosaur. Students are typically given only this information - leading them to the conclusion that it was a dinosaur in the process of turning into a bird.[22] Share these additional facts with your students (which are left out of textbooks):

- This fossil had fully formed wings, feathers, and in every way would have been classified as a fully functional bird if found alive today.

- Often in textbooks, Archaeopteryx has been classified as part bird/part dinosaur because it had teeth, claws on it wings and a tail. Today we have birds with claws on their wings: the Hoatzin. The Hoatzin chick with claws on its wings climbs around rain forest trees like a monkey. Ostriches also have finger-like claws on their wings. No living birds have teeth, but other fossilized birds such as Hesperornis, an aquatic bird, did have teeth – yet it was fully a bird.

- Think about this – Archaeopteryx is routinely presented in textbooks as the link between reptiles and the first bird. Yet fully formed birds, essentially indistinguishable from modern birds, have been found in <u>lower</u> rock layers. If an essentially modern bird is found in rock layers lower than Archaeopteryx, then this fossil can't be the ancestor to all birds! So why is it still being used as evidence for evolution in textbooks?

There are even more serious problems with the fanciful belief that a dinosaur could turn into a bird. For a dinosaur to evolve into a bird would require major changes.[23] Consider a few:

- **Bones** - Solid dinosaur bones becoming lighter and hollow like bird bones.
- **Respiratory system** - Reptiles, which are like dinosaurs, have lungs like us, a bellow system to move air in and out. Bird lungs have seven to nine air sacs in which air flows in one direction Even the bird's hollow legs are involved as oxygen flows through. How would the bellows style lung evolve gradually into a bird's air sacs? In the intermediate stages, the poor animal would die.
- **Skin to feathers** - scales are just folds in the skin; feathers are complex structures with barbs, barbules and hooks. Also, scales and feathers are made from completely different biochemicals.
- **Cold-blooded to warm-blooded**

These are just some of the changes needed for a dinosaur to evolve into a bird. Evolutionists have not found the transitional creature in the fossil record, neither have they explained how it would happen. They do not go into the details. And by the way, in each one of the cases listed, how does the creature survive with half the characteristics of each creature? The belief in dinosaur to bird evolution is promoted not because of evidence supporting it, but in spite of evidence showing that it is an impossibility. Why? Because the only other alternative, the creation of very separate types of creature (birds and dinosaurs), is ignored.

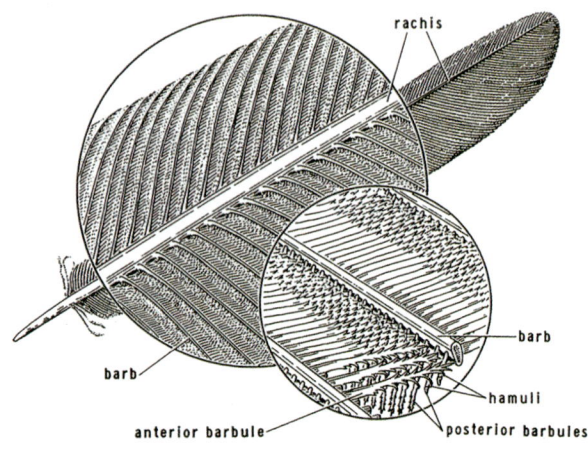
© University of Michigan, Museum of Zoologyv

Dinosaurs and Dragons

KEY POINTS TO REMEMBER:

1. Fossils of supposed "evolving birds" (like Archaeopteryx) have characteristics of birds still alive today.
2. Fossils of birds have fully formed feathers and wings.
3. Major internal changes needed to transform a dinosaur to a bird have never been explained or found in the fossil

Student Activity - Evolution of a New Animal

If you could find a Lego set of an animal (a dinosaur would be great) have one group of students move one block at a time and try to transform it into a completely different animal. They can use blocks from a different set if they wish. If Legos are not available, maybe some other set of blocks that form a complex structure.

+ + + + + + + +

> There is one unbreakable rule as the students try to transform one structure (animal) into another:
> With each block added or removed, the animal must still be able to function and/or survive

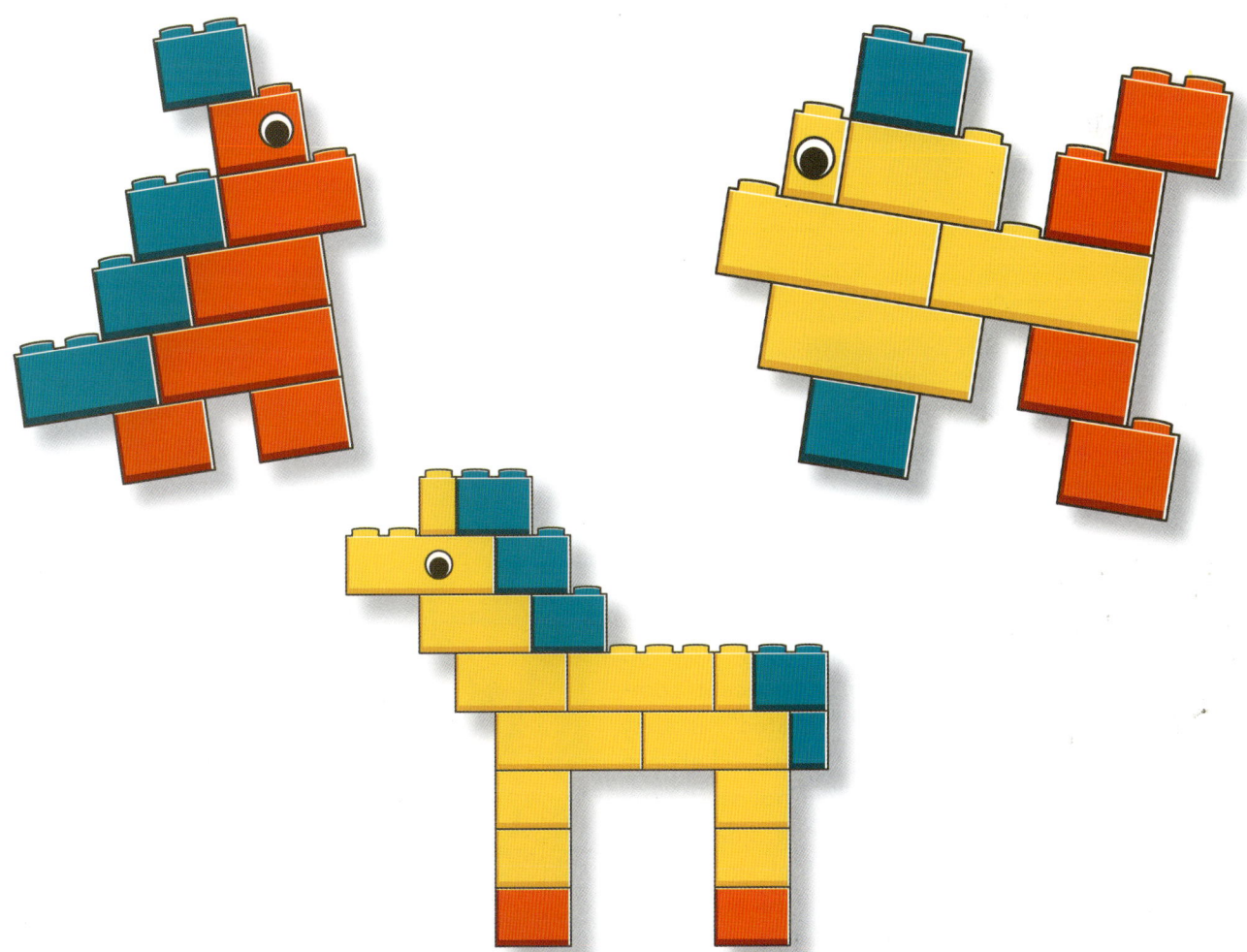

Have other group of student judge whether each change allows the animal to survive as it is transformed. Even one useless or non-functioning part dooms it to extinction – no evolutionary advancement!

Have the students doing the transformation try to justify how the slowly transforming creature could survive.

They will quickly see that transforming one creature to another by **evolution simply does not work.**

E. Is There 98% Similarity Between Human and Chimpanzee DNA?

The short answer is no. Initially, scientists thought so, but as the genome of chimps and humans are furthered studied, evidence is found that chimps and humans have differences. [24] Here are some research discoveries showing the differences between chimp and human DNA:

1. In 2005, scientists discovered that the chimp genome is 12 % larger than the human genome.
2. In 2003, there was a calculated 13.3% difference in sections of our immune systems.
3. There is a 17.4% difference in the gene expression in the cerebral cortex.
4. Chimps have 24 chromosomes and humans have 23 chromosomes. Initially, evolutionists thought chromosome #2 fused together on the chimp to create the human 23 chromosomes in humans. However, with further studies, the telomers (the tips of the chromosomes) are different, and they shouldn't be if they had been fused. Also, the center centromere, the pinch point, is very different. [24a]

Imaginary Evolutionary Tree

Now let's take a look at how they sequenced the chimp genome. Researchers assuming chimps and humans have a recent common ancestor "cheated" as they analyzed the similarity between the two. They did not start from scratch by comparing, side by side, the letters (nucleotides) of each, but took small pieces of chimp DNA and aligned them with the human genome. Then the human genome was removed, leaving a pseudo-chimp genome. In other words, the human genome was used as a scaffold for the chimp DNA. It is a trick which starts by assuming the two are similar, and then does not even consider areas where they are vastly different.

In 2012, Dr. Jeffrey Tomkins and Jerry Bergman reviewed the published studies comparing human and chimp DNA. When all the DNA was taken into account and not just pre-selected parts, they found, "it is safe to conclude that human-chimp genome similarity is not more than ~87% identical, and possibly not higher than 81%." Since the DNA of humans and chimps are approximately 3 billion letters long, this means there are actually about 600,000,000 differences! With these numbers, now evolution is ruled out.[24b] This would be analogous to taking a huge book and randomly changing one out of every five letters in the entire book. Not only would it become unreadable, but we all know the book would not change into a better book!

Y Chromosome Difference Between Man and Chimps

	Human	Chimp	Difference
Distinct Genes	693	~455	33%
Protien coding elements	88 million	41 million	53%
No corresponding counterpart chromosome	30%	30%	30%

<u>VAST</u> Differnce between man and monkey!

So why was it repeated that we are 98% similar to chimps for so long? Because of the faulty assumption that human evolution is a fact and because of leaving a Creator out of their thinking. Thus, only the similar parts of the DNA is examined - instead of looking at the entire organism. This would be similar to comparing an old original model-T automobile to an expensive modern sports car. If you only listed the similarities, you might assume one "turned into" the other. But in reality, both exist because they have a common designer.

Both have tires, headlights, wind shields, wipers, hoods, roofs, seats, steering wheels, pistons, engines, use gasoline/oil, door handles, etc. Depending on the features selected, they are 98% identical, yet it would be absurd to believe random changes could transform one into the other.

KEY POINTS TO REMEMBER:

1. Human and chimpanzee DNA is at most 80% identical, not 98% identical.
2. There are hundreds of millions of differences, between apes and man, the apes could never have survived randomly changing its DNA, an improved human DNA would never result.
3. Just like a book could never slowly transform itself into a different book by random changes, the DNA of one animal could not transform one animal into a different one by random changes.

Photo Credit: Wikipedia - ModelTMi and Freepik

Student Activity - Similarity Proves Evolution?

Have your students select two similar objects and then list all the similarities and differences. Maybe break them into groups. Have them calculate the percentage similarity.

+ + + + + + + +

[(Number of similarities) / (total number of similarities + differences)] X 100 = % Similarity

When done have them present their results and then ask these questions:

- Which objects chosen were most similar?
- Which objects chosen were least similar?
- Of all the objects chosen, does a higher percentage of similarity prove that one thing turned into the other thing all by itself (i.e. by random evolutionary change?)

Is the more logical answer that both objects were designed by intelligence?

+ + + + + + +

+ + + + + + +

B. Have Fossils Formed Over Millions of Years?

a) Have fossils formed over millions of years?

The rock layers of the Earth are assumed to have been laid down over billions of years. These time frames are required if the belief in evolution is to be maintained, yet how can we KNOW the rock layers of the earth took this long to form? The other worldview, from the Bible, is that the Earth is less than 10,000 years old, and approximately 4,500 years ago there was a worldwide flood which laid down the sedimentary rock layers, resulting in the formation of the fossils found in these rock layers.

Sedimentary rocks (sandstone, limestone, shale, coal, mudstone, etc.) all formed underwater and cover 75% of all exposed land surfaces. There are places on earth where sedimentary rock extends down 40,000 feet. These rock layers are very much like a stack of plates sitting on a kitchen counter. If you walked into someone's house and see this stack of dishes, you know the bottom dish was placed there first, but you may not know how long ago this happened. The entire stack could have been placed in position rapidly in a matter of seconds, or the bottom dish may have been there for weeks while the top dish was just placed in that position minutes earlier. The same is true of the sedimentary rock layers of the Earth and the fossils they contain. It is assumed, not a scientific fact, that the lowest layers are billions of years old. These layers can extend deep into the earth, because during the flood of Noah, enormous waves filled with ground-up sediment rapidly settled into low depressions in the Earth, filling them with layer after layer of loose sediment, which subsequently turned into rock layers filled with fossils.[25]

Noah's Flood and Geology

Think about this - what happens to a dead animal? Scavengers eat it; bugs and bacteria cause it to rot and decay, eventually leaving no remains behind. It takes very special conditions in order to make a fossil. Here is the recipe required to create a fossil:

1. Fast coverage by sediment. Scavengers and bacteria don't eat it.
2. Deep coverage by sediment so no oxygen is present to start decay.
3. Lots and lots of water so the minerals can seep into the bone and turn it into stone.

Scientists have also found that it does not take millions of years to make petrified wood. Five Japanese scientists studied a small lake cradled in the crater of the Tateyama Volcano in central Japan. The crater is filled with steaming acidic waters which gushed from the bottom of the lake. This mineral rich solution filled a 35-foot pond with a waterfall that cascaded over the edge. A piece of fallen wood in the overflow was found by the scientists to be petrified with the mineral silica. This petrified wood was only 36-years-old. These scientists then experimented by fastening pieces of fresh wood on wires and lowered them into the lake. After seven years, it was found to be petrified, or turned to stone. It did not take millions of years to make petrified wood, but rather the right conditions. The Flood of Noah's time would have created the perfect conditions for petrifying trees rapidly - plenty of mineral rich waters and volcanic activity.[26]

All scientists have the same rocks and fossils to examine, but if the worldwide flood is eliminated from their thinking, they are guaranteed to misinterpret the past. Fossils do not prove evolution, and they did not form over millions of years.

KEY POINTS TO REMEMBER:

1. Fossils do not take millions of years to form.
2. Fossils can only form if dead organisms are buried deeply, rapidly and surrounded by lots of mineral-filled water. We seldom see these conditions happening today.
3. The Flood of Noah explains geology and fossils better than evolution and enormous time periods.

Student Demonstration - Are Fossils Forming Today?

Teachers need to acknowledge what textbooks teach about the slow formation of the Earth's rock layers and fossils they contain so that students can pass standardized tests - but they need to also share the second possibility, i.e., that a year-long world restructuring flood ripped up the entire surface of the Earth, sorted/redistributed sediment across the globe, and buried trillions of organisms, which rapidly and relatively recently turned these dead organisms into fossils. Ask your students the following questions so they can sort out for themselves which is true:

- When a tree dies in the forest does it turn into a solid rock fossil?
- When a fish dies in the ocean does it drop to the bottom and turn into a rock?
- Put a dead minnow in a glass of water, does it sink? It bloats and floats. Next bacteria and other animals eat it. A few bones will drop to the bottom but even these will be consumed. Do you see the bottom of lakes and oceans full of dead creatures? When a fox dies in the forest will its bones turn into solid rocks over time?
- If you bury your dog and it was dug up 100 years later, do you think the bones would still be there, turned to rock?

Explain that for a fossil to form, something has to be buried extremely fast and deep. Otherwise, scavengers, fungus and bacteria will destroy it rapidly - within months or years. Lots of water filled with minerals must also be moving through the dead organism's cells. Fossils are essentially not forming today, yet the rock layers all over the earth are filled with trillions of fossils. It has also been shown that mineralization (the replacement of organic tissues with silica and minerals) can happen within months under the right conditions, and does not take millions of years. Something very different happened in the past for fossils to fill the rock layers of the Earth.

After explaining both models for the origin of rocks and fossils, ask your students which is the best fit for the evidence we find:

- The flood of Noah?
- Billions of years of sediments slowly covering the dead creature or plant?

The following videos give exciting evidence for the reality of a worldwide flood that created the rock layers of the earth:

b) Does the Cambrian Explosion represent rapid Evolutionary development?

In the study of earth's geological history, textbooks, TV programs and museums put twelve systems in a vertical order and call it the geological column. In this geologic column, the rock layers or strata form a sequence from bottom to top. The complete geological column is not found at any location on earth, it is only found in diagrams! During the 1600s and 1700s, the column was widely accepted to represent the order of what was buried during Noah's flood; first the sea creatures and eventually the land creatures. During the late 1700's and 1800's, the Bible became increasingly ignored, evolution became a popular belief, and the interpretation of the geological column changed. The sequence of fossils in the rock layers is now assumed to represent the evolution of one-celled creatures (such as bacteria) to complex animals over millions of years.

But what do we find in the fossil record? Near the very bottom of the geological column is the Cambrian layer. In the Cambrian layer, we find fossil remains of complex creatures from all the major animal groups. In fact, we find representatives of every major animal phylum in existence today, and phyla that are now extinct. Phylum is the most basic body structure into which organisms are classified. There are twelve phylum and everything with a backbone is one of the 32 groups. This sudden appearance of complex life is called the "Cambrian Explosion." The Cambrian Explosion is actually extremely compelling evidence against slow and gradual evolution! When we study the facts, we find God's Word is true. In the beginning, He created the plants and animals in six days. When the Flood of Noah took place, all of these animal types died and were buried in sediment that turned to stone. In general, sea creatures would have been buried deepest and land creatures later and higher up.

Photo Credit: Mesa Shumacher/Santa Fe Institute

The Cambrian Explosion testifies to Scripture – God created every kind of animal, fully functional, the first week of creation. This is exactly what we find in the Cambrian rocks.

As we look at the rock layers of the earth, one of the most striking features is known as the "Great Unconformity." This is a scoured-off surface is generally between metamorphic rock (basalt, granite, etc.) and sedimentary rocks (laid down under water). Textbooks either ignore this feature or mention it as if it is a fact of science that the junction between these surfaces represents 500 million years of "missing time." In other words, it is assumed that 500 million years passed while life was evolving and no rocks or evidence was left upon the Earth for these events or this passing time.[28] Evolutionists are acknowledging that some big event happened that scoured-off the surface of the earth. In their worldview, they cannot acknowledge the Flood because Noah's Flood eliminates the time needed to leave God out and believe in evolution. Yet, what would make a <u>worldwide</u> "Great Unconformity?" And what do we find immediately after the Great unconformity – the Cambrian rock layer. These Cambrian rocks would have been the first ones to form several months into the year-long flood of Noah. Sea life would have been captured within this newly forming rock layer as flood waters, filled with sediment which had been scoured off the land surface of the earth, This would have encapsulated every basic body structure of the previously created sea life. There are no missing 600 million years of Earth history. These sea creatures did not evolve first, nor were there early "primitive" forms of life - they were simply the wide variety of ocean life first buried during the Flood.

Textbooks mistakenly state that evolution caused a sudden explosion of life upon the earth about 600 million years ago. This is faith, not science, and it is unsupported by any scientific observation. The fossil evidence actually falsifies evolution and testifies to the sudden explosion of life. What the evidence shows is the sudden appearance of completely different forms of life - fully formed and completely functional.

Student Activity - Are Fossils Forming Today?

Ask your students this series of questions about the Cambrian explosion of life to help them separate the evolutionary beliefs they are reading in their textbooks from the actual observations of science:

+ + + + + + + +

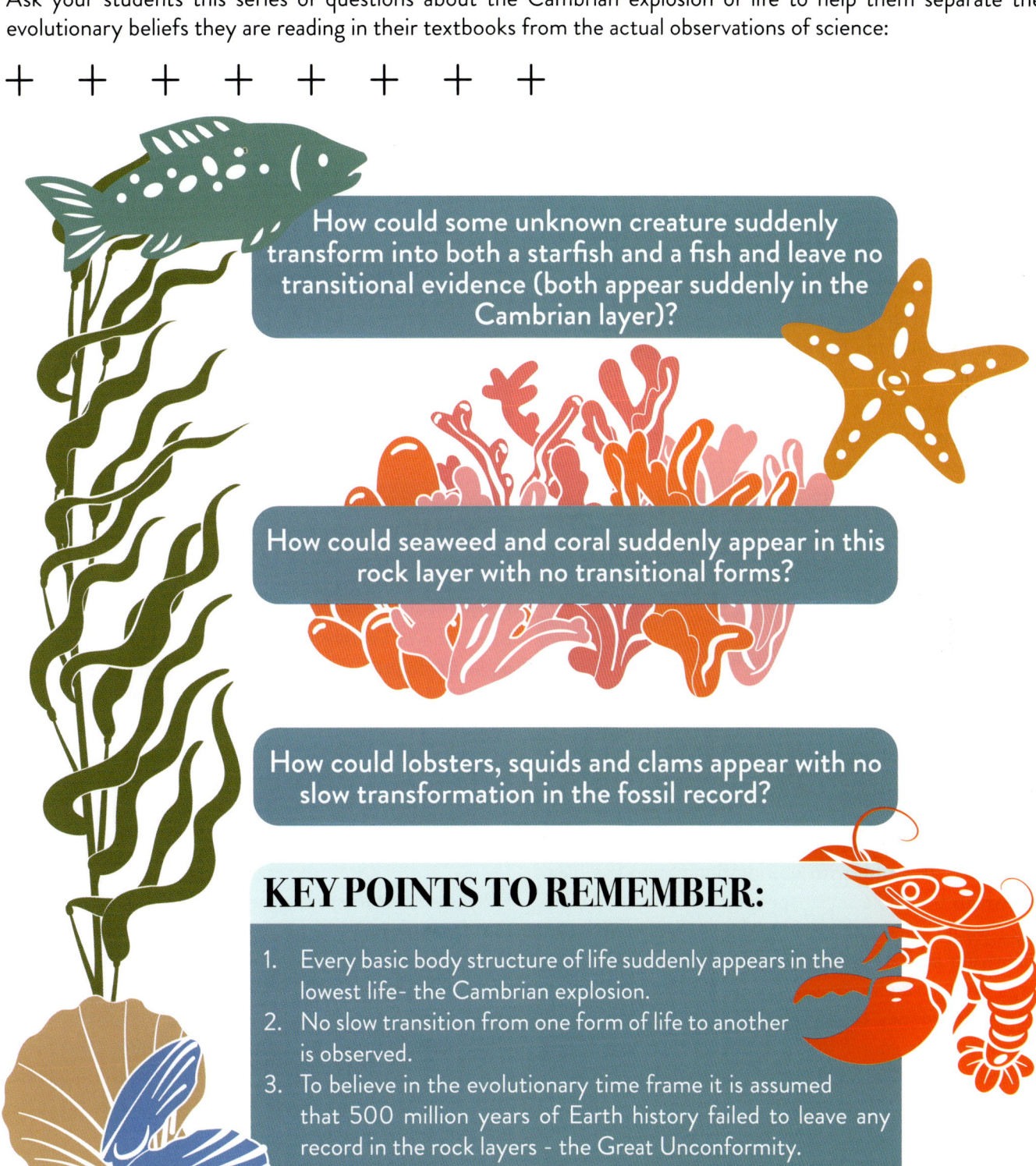

How could some unknown creature suddenly transform into both a starfish and a fish and leave no transitional evidence (both appear suddenly in the Cambrian layer)?

How could seaweed and coral suddenly appear in this rock layer with no transitional forms?

How could lobsters, squids and clams appear with no slow transformation in the fossil record?

KEY POINTS TO REMEMBER:

1. Every basic body structure of life suddenly appears in the lowest life- the Cambrian explosion.
2. No slow transition from one form of life to another is observed.
3. To believe in the evolutionary time frame it is assumed that 500 million years of Earth history failed to leave any record in the rock layers - the Great Unconformity.

C. Does the Sequence of the Fossils in the Rock Layers Prove Evolution? There is a general tendency to find "simple" sea life lowest in the rock layers (clams, trilobites, sponges, nautiloids, etc.), more mobile sea creatures higher up (fish, sharks, etc.), amphibians, reptiles, and land plants higher still, followed by mammals, birds, and finally mankind. This is shown in textbooks and every natural history museum in the world as "proof" that one animal evolved into another over huge periods of time. The various levels of rock layers have been given scientific names corresponding to eras of earth history. In actuality, this complete column of rock layers with the fossils they contain, can be found nowhere on Earth. The Grand Canyon has the most, but still only contains about half the supposed geological strata of evolutionary history. The geological column is simply a concept to fit history into a slowly evolving storyline for life.

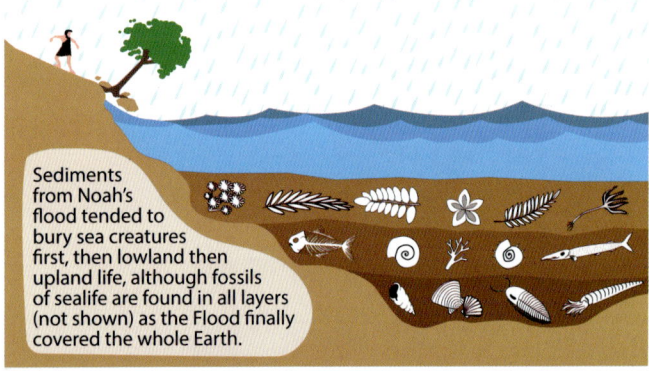

Sediments from Noah's flood tended to bury sea creatures first, then lowland then upland life, although fossils of sealife are found in all layers (not shown) as the Flood finally covered the whole Earth.

The problem is again leaving out the other logical alternative. A world restructuring flood upon the planet, which the Bible says happened about 4,500 years ago, would have pulverized everything on Earth and redistributed the sediments filled with trillions of dead organisms. This world pulverizing flood would have a general tendency to bury things regionally along with other animals found in their ecological living zones. Bacteria and microorganisms outnumber large multicellular organisms by orders of magnitude, and have always been found in every possible layer of rock. The fact that they have been found in layers of rock below other forms of life simply means they worked their way into those locations before flood waters buried larger creatures at higher levels.[27] When looking at the fossil record as a whole, what do we find?[27a]

95% are marine invertebrates, mostly shellfish
<5% are plants such as algae and trees
<1% are vertebrates of all kinds.

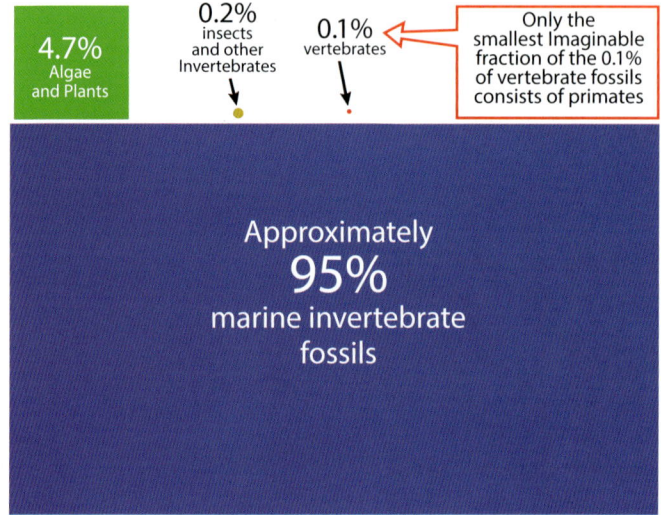

Most fossils are marine invertebrates, those without a backbone or vertebra. This would include clams, coral, and trilobites. Of those remaining, five percent are mostly plants. Less than one percent of all fossils are fish, and even fewer are land animals such as dinosaurs, amphibians, mammals and birds. The number of dinosaur skeletons found are currently estimated to be around 2,000 "good skeletons," meaning not just a single bone.

Why are most fossils marine invertebrates? During a worldwide flood, the ones buried first would have been the ocean invertebrates. As the tsunamis came on land, they carried with them marine creatures. Meanwhile, the land creatures were moving to higher ground and succumbed to the flood waters where they bloat and float. The Flood waters could have swirled them into areas where they were covered by sediments, creating dinosaur graveyards.

The fossil record is best understood as a result of the Genesis Flood. Things living at the bottom of the ocean (clams, trilobites, sponges, nautiloids, etc.) would be buried first (and therefore deepest). 95% of all fossils are ocean creatures. Their bodies would have been buried amidst rapidly accumulating sediment. Slower-moving land animals (such as amphibians and reptiles) would be buried higher up, dinosaurs would likely congregate and be buried in distinct layers, and the most mobile creatures, such as mammals and birds, would be buried last (and therefore highest). Mankind would likely be the last to die, and very few, if any, human skeleton fossils would form because their bodies decayed rather than being buried in sediment.

Evolutionists deny a worldwide flood. This results in a misinterpretation of the fossil record, rock layers, and the age of the earth.

KEY POINTS TO REMEMBER:

1. The entire geological column exists nowhere on Earth – it is simply a visual tool used to promote evolution.
2. Most fossils are sea creatures, and they are often found buried with land animals. This happened during the Flood of Noah.
3. The geological column represents the order of burial during the Flood - not billions of years of Earth history.

D. Does Pangaea (Plate Tectonics) Prove the evolutionary Time-frame?

Almost every earth science and world history book start the story of earth's history by showing a picture of the ancient earth with all of the continents squeezed together into a single super continent called "Pangaea." We currently measure the continents moving apart at about 2.5 cm /yr., or about as fast as your fingernails grow. Extrapolating back in time, at this rate it would take hundreds of millions of years for the continents to reach their current positions.

When the other worldview is taken into account (that there was a worldwide Flood), it becomes obvious that there has not always been a slow continental drift, but the past shows a "continental sprint." During the yearlong Flood, the plates would have moved quickly, ramming into each other and creating the current mountain ranges of the world at the end of this Flood. Creation scientist and plate tectonic expert Dr. John Baumgardner, while working at Los Alamos National Laboratory, created the mathematical model which is used to explain continental movements. He determined that the continents sprinted, not drifted. He has shown through laboratory experiments and modeling that the underlying rock layers of the earth would undergo shear thinning during a major earth upset, and the viscosity (thickness) of this underlying rock layer could change by a factor of ten billion as the continents started to rapidly move. This work indicates that the continents could have moved into their current positions in a matter of months during Noah's Flood, not over hundreds of millions of years.[30,31] Once again, when the Flood is ignored, wrong conclusions about the earth's past result.

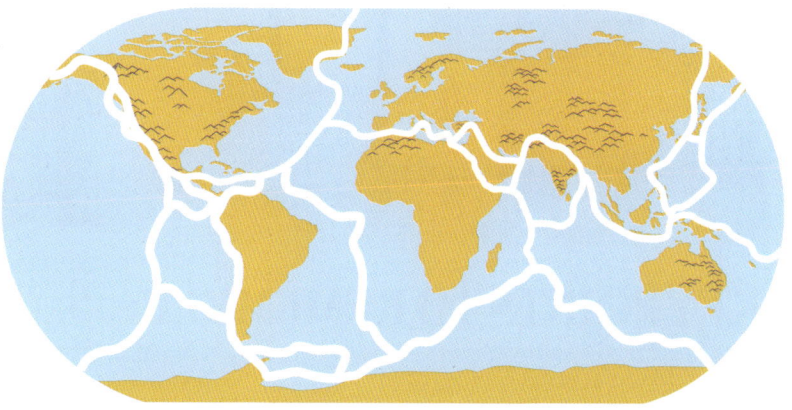

Student Activity - How Long For Continents to Move?

Have students cut out outlines of the world's continental plates as shown in most textbook pictures of Pangaea. Once cut out, have them group them into one mass and then move them quickly to today's positions. Explain how the Indian plate slamming into the Asian plate would have created the Himalayas. Did you know the Himalayas are still rising? The Himalayan Mountains are still growing higher, at a rate of about 2.4 in/year (6.1 cm/year).

+ + + + + + + +

Have more advanced students calculate the time required to move into their current positions based on starting assumptions:

> It has moved 4500 KM at the current rate of 2.5 CM per year
> (4500KM)1YR/2.5CM)(100CM/M)(1000M/KM)=180,000,000 Years
> It has moved 4500 KM during the Flood at a shear thinning rate 1 billion times faster
> 180,000,000/1,000,000,000 = (0.18 years)(12 month/year) = 2.16 months!

How can we **KNOW** for sure at which rate the movement happened?

Use this demonstration:

- Put two children's trains on a single track facing each other. Tell students these trains represent the continental plates moving toward each other.

- Now very slowly push them together so they barely touch. Did they bunch up into a mountain of rubble?

- Next fling each train rapidly at the other so they crash at high speed. A heap or pile of train parts results, i.e. a mountain of rubble.

This visual demonstration will be remembered by students as explaining that the mountains formed rapidly in the later stages of the Flood of Noah, not slowly over millions of years.

KEY POINTS TO REMEMBER: + + + + + + +

1. It is continental sprint, not continental drift.
2. There are valid scientific reasons to believe it happened within months during the Flood of Noah.

E. Do "Living Fossils" Support Evolution or Creation? Living fossils are fossil that are essentially identical to their living counterparts today. If evolution is true we should find one creature evolving into another. What do we observe? Fossil starfish purportedly dating more than 500-million-year-old look essentially like today's starfish, no evolution. A maple leaf fossil looks like today's maple leaf, no evolution. The fossil Polistes wasp looks like today's Polistes wasp, no evolution. Bacteria are considered the oldest creatures on earth, supposedly appearing about 3.5 billion years ago. Yet, exactly the same types and structures of bacteria have been identified in fossils as we find living today. In other words, known types of bacteria have SUPPOSEDLY not changed in any significant way for 3,500,000,000 years, while over the same time period, other bacteria evolved into trees, whales, dinosaurs, birds, monkeys, and people!

Dr. Carl Werner has published a book, **Living Fossils**, documenting not just a few, but hundreds of plants, animals, and sea life which show no significant difference between fossils and the same creatures still living today.[34] This section contains a few examples of "living fossils." Remember that the dates associated with the fossil version is not a fact of science, but an assumption based on ignoring the evidence that a worldwide flood buried these organisms less than 4,500 years ago.[32,33]

We find hundreds of different types of "living fossils" which show us that plants and animals have stayed essentially the same throughout time. Evolutionists cannot explain why certain creatures supposedly evolved while others remained unchanged for hundreds of millions of years. They simply call this lack of change "stasis," assume that evolution happens at different rates, and teach that no matter what they find in the fossil record (lack of change or vast differences between organisms) it always supports evolution.

Ginkgo Leaf – no change in a purported 400,000,000 years!

Dragonfly – no change in a purported 300,000,000 years!

Ferns – no change in a purported 360,000,000 years!

Nautilus – no change in a purported 400,000,000 years!

Horseshoe Crab – no change in a purported 450,000,000 years!

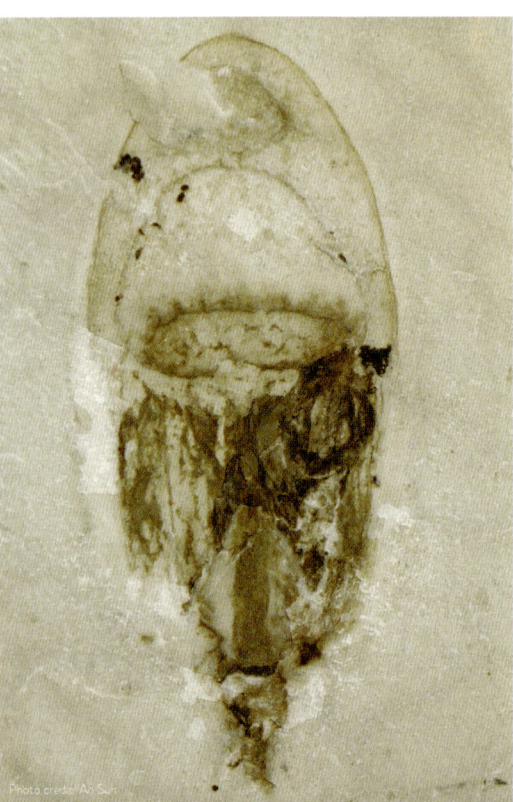

Jellyfish – no change in a purported 500,000,000 years!

KEY POINTS TO REMEMBER:

1. Living fossils are fossils of plants and animals that show no significant change in basic body structure. This is called stasis and is evidence for creation, not evolution.

F. Did the Grand Canyon formed over millions of years? Six million people each year visit the Grand Canyon. It is the most visited natural wonder in the world. Every one of these visitors is told that this great chasm cut into the earth was created slowly over millions of years as the Colorado River slowly eroded away the rock layers to create the canyon. This explanation is repeated in textbooks around the world as a fact of science, and no alternative is mentioned. The Grand Canyon extends down 1,800 meters (6,000 feet) through a dozen different (Noah's Flood created) strata layers all the way into original creation week rock. But when and how did the Grand Canyon come into existence? All of the park signs and textbooks tell us that the Colorado River carved out the Grand Canyon over millions of years of slow erosional processes. However, there are enormous problems with this explanation.

The Grand Canyon is 277 miles long, about a mile deep, and four to eighteen miles from rim to rim. *Geological facts to ponder:*

1. Where's the dirt? Simple calculations have 1,000 cubic miles eroded away. Normally the sediments will be at the delta of a river. Look at the Colorado River's mouth, there is virtually no delta! The little sediment found at the delta represent only thousands of years of erosion, not the evolutionary millions of years. So where is the dirt? It had to be removed catastrophically.
2. Stable cliffs. The cliffs of limestone and sandstone are stable. The dark color is the coating of desert varnish, which forms slowly over many years. The fact that you see the desert varnish indicates stable cliffs with not much rock fall. The cliffs are not experiencing slow erosion. The cliffs testify to recent catastrophic erosion and are now stabilized.
3. No talus. Talus or debris is lacking at the bottom of the cliffs. If the canyon is millions of years old, there should be vast amounts of talus or rock debris at the base of the cliffs. Now check the side canyons ending in the U-shaped amphitheaters, do you see talus? NO! It is clean. Some amphitheaters are hundreds of feet deep and extend back a mile from the river. Most do not have a water source to remove the possible talus. If you believe the canyon was slowly eroded, where is the talus and what formed the side canyon which has no river flow? The lack of talus affirms a recent and catastrophic event.

These geological facts give testimony to a recent and catastrophic event, not millions of years of slow erosion. The Grand Canyon was carved rapidly with lots of water in a short amount of time, thus no delta, no talus, and stable cliffs. The Grand Canyon is a testimony to a recent catastrophe![35, 35a]

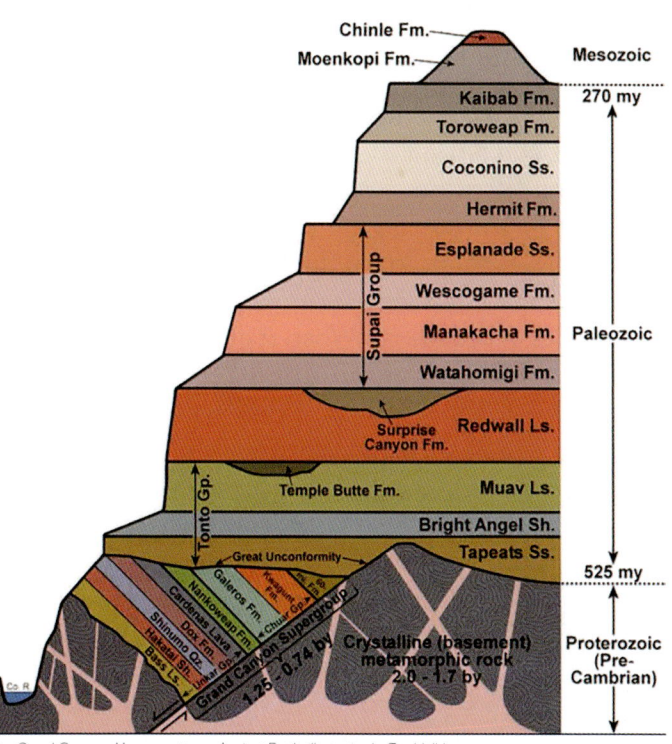

The Grand Canyon, Monument to an Ancient Earth, illustration by Tim Helble

How could the Flood of Noah have carved out the Grand Canyon rapidly?

One of the most baffling geological features of the Grand Canyon is the canyon cutting through a great plateau and not around the plateau. The Grand Canyon should not be where it is. The Colorado River runs south, then it abruptly turns 90 degrees and into the heart of the uplifted Kaibab Plateau, which is 3,000 feet above. Evolutionists believe that the Grand Canyon was carved over millions of years by the Colorado river. What is the Biblical perspective? During the Flood of Noah, sedimentary layers were laid down. At the end of the Flood, Psalm 104:8 speaks of the mountains rising and the valleys sinking. As the continents rose, the Flood water drained off into the new ocean basins. Massive sheet erosion occurred while continents were being lifted up. Sheet erosion is massive water movement removing an even amount of sediment off the surface of an enormous area of land. Sheet erosion just north of the Grand Canyon carved the Grand Staircase, a sequence of ascending cliffs, leaving behind the colorful cliffs of Zion and Bryce Canyon National parks. In this area alone, it has been estimated that this sheet-like erosion eroded 100,000 cubic miles of sediments. When standing at the Grand Canyon rim, imagine one mile of dirt above you.

Near the end of the continent rising and the draining of water into the oceans, oversized river valleys formed around the world. In the Grand Canyon area, as land neared the surface, the rapidly moving water cut deeply into the newly laid sediment. Another possibility is that Flood waters became trapped and large lakes formed to the north and east of the Grand Canyon. These lakes contained water estimated to be three times the volume of Lake Michigan! The impounded lakes would continue to increase in size. Meanwhile, the natural dam started to weaken and was soon breached. Now the lakes emptied, and the water rushed forward exploiting any channels already carved. The Grand Canyon was formed. From a creation geologists' viewpoint, the Grand Canyon was carved by a lot of water in a short time period and there is plenty of evidence to support this interpretation.[35b] No scientist was present to see the Grand Canyon form but the evidence for a <u>rapid</u> catastrophic formation is compelling. It is simply ignored by geologists trained to think in terms of slow, gradual evolutionary change.

The Grand Canyon

KEY POINTS TO REMEMBER:

1. No scientist was present to see the Grand Canyon form.
2. There is compelling evidence that it formed recently and rapidly at the end of Noah's Flood.

G. Did Dinosaurs Go Extinct Because of a Meteorite Impact?

Certain rock layers of the earth are filled with fossils of animals which no longer exist. These animals are assumed to have died during some major extinction event which happened in the past. A classic example is the disappearance of the dinosaurs. Textbooks tell students that dinosaurs went extinct because a huge asteroid hit the earth about 65 million years ago wiping out dinosaurs. Why would such a catastrophe destroy every single type of every dinosaur yet leave other creatures alive to flourish and fill the earth?

The purported "evidence" supporting this dinosaur extinction event was the discovery of iridium in the K-T boundary or Cretaceous-Tertiary. How did this high concentration of iridium get between these two layers? It has been taught that an enormous asteroid brought the iridium, and it was spread around the earth. This is taught as a fact of the past, but no-one observed this event, and it is just assumed to have happened. All dinosaur fossils, found on every continent, are buried in sedimentary rock layers that formed under water. Land dinosaurs are often found buried with fossils of sea creatures. This testifies to burial by flood waters, not a meteor impact.

An alternative possibility is that this iridium came from deep within the earth during major volcanic eruptions which were happening during Noah's Flood.[36,37] During the year-long Flood, volcanoes were erupting bringing the iridium to the planet's surface. The iridium would have mixed with the water and been distributed and buried in a distinct layer. Meanwhile, dinosaurs were caught in the Flood waters and buried in the same general layer that captured the iridium release from the volcano.

Brain and Sleep?   Banning Water

KEY POINTS TO REMEMBER:

1. Dinosaur bones are always found in sediments laid down under water.
2. It was Noah's Flood, not an asteroid, which buried dinosaurs and created their fossils around the world.
3. Iridium also comes from volcanic eruptions on Earth.

SECTION IV - PROBLEMS WITH THE MECHANISMS OF EVOLUTION

A. Natural Selection

When Charles Darwin published *The Origin of Species* in 1859, he brought what was thought to be scientific credibility to the concept that man developed slowly from previous life forms. The intellectual community of Northern Europe was ripe for a naturalistic explanation of life. The distortion of Biblical Christianity was bringing faith in the supernatural under increasing ridicule, and humanism (man making himself the center of all things) was rapidly replacing the Christian belief in absolute truth. Thus, when an alternative to creation seemed to have been found, it was rapidly accepted as fact. The easiest way to reject the authority of a Creator is to remove the belief in the existence of that Creator.

In actuality, Darwin proved neither where life came from nor how it developed. Darwin proposed several things concerning the origin of the variety of life on our planet. The first was that "the species are not immutable" (i.e. we came from a previous simpler form of life). Darwin's second proposal was that this transformation of one life form into another was driven by a process called natural selection (popularized as "survival of the fittest.") His third proposal was the rejection of a Biblical time frame and a worldwide flood as an explanation for the geology of the planet. This allowed huge time periods supposedly needed for the evolutionary transformation of bacteria to people. Believers in evolution still assume that given enough time, there is essentially no limit to evolution. But can this seemingly magical force transform amoebas into man? It is pointed out by believers in evolution that some mutations find a useful purpose. (We will cover mutations in more detail in the next section.) Therefore, it is taught that even a slightly advantageous change will be passed onto the next generation, which, exceedingly slowly, transforms one type of creature to another. This process is known as natural selection and will sound logical if only presented in this way. But how does natural selection hold up under the light of scientific scrutiny as the directing force behind life's origin and diversity? Dr. John Sanford, retired Cornell University geneticist, inventor of the "gene gun" and 25 other patented discoveries, uses the following illustration to expose the fallacy of this evolutionary belief.[38]

Dr. Sanford likens the amount of information required to transform a single-celled organism into a human as far greater than the information required to transform the manufacturing factory for a child's little toy wagon into the most complex form of transportation known to mankind - a space vehicle. Can natural selection, acting on accidental changes to the assembly directions of the little wagon, accomplish this transformation? The opposite is actually true because natural selection cannot even prevent the deterioration of the wagon, let alone increase its complexity.

Natural selection is similar to the quality control department at the wagon assembly plant and our genetic code is similar to a document containing the entire manufacturing process for the little wagon. Everything needed to manufacture the wagon, including the specifications for all of the materials of construction, all of the individual components, the processes needed to manufacture them, all of the metal press specifications, all of the robotics and programming language, the assembly instructions, the paint specifications, the employee benefits manual, i.e. everything needed for the wagon's construction, is attached as a manual below the wagon. The next wagon to be produced must use only the information in the existing wagon to make the next copy. The quality department (natural selection) can only see the finished wagon, not the enormous amount of information in the manufacturing manual (the genetic code of the wagon). The question is: can random changes in the assembly manual (the genetic code) allow the quality control department (natural selection) to transform the little red wagon into a better wagon and ultimately into a spaceship?

$$+ \quad + \quad + \quad + \quad + \quad +$$

$$+ \quad + \quad + \quad + \quad + \quad +$$

The amount of information contained within the simplest single cell organism (similar to the information required to build a wagon assembly factory) would fill a small library. Suppose you started with a perfect set of instructions in this library and randomly changed a few dozen individual letters throughout the instructions. Very few of these changes would be critical for assembly or cause a faulty wagon - which the quality control department (natural selection) would reject. It is far more likely that just a few random changes to an entire book (these are called mutations in living organisms) would result in no noticeable change and the wagons would roll off the assembly line with mistakes in their manuals intact -- to be used in the creation of the next generation of wagons. This next generation would then have another dozen barely noticeable mistakes added. Given enough generations of the wagons, there would be ever-increasing, letter-by-letter mistakes in their assembly manuals; eventually the point would be reached when wagons could no longer be produced from the instructions because there were so many tiny mistakes present.

Large mistakes can be eliminated by natural selection, but not the small mistakes because, one wrong "letter" at a time, they are essentially undetectable in the final product. Yet in the end, they will drive the manufacturing process to extinction the same way one rust molecule at a time will eventually destroy a car. This is exactly what is happening to the human genome at an alarming rate. Hundreds of tiny mistakes are building up with each generation. Natural selection could NEVER transform one creature into a more complex creature, and it cannot even stop the current deterioration of life.[39]

Design Testifies To Creation of Life

Textbooks promote examples of evolution which are caused by what they call "useful" or "positive" mutations. This would be like randomly changing the word "red" to blue" in the painting section of the manual. That wagon may sell very well in the market. But remember, no increase in the amount of information resulted and the blue wagon takes with it all the tiny detrimental mistakes which will continue to grow. Now let's look at the misleading examples of natural selection used in textbooks to convince students that bacteria-to-human evolution is true.

KEY POINTS TO REMEMBER:

1. Natural selection cannot add new information to the DNA code.
2. Every generation has more random changes to its coded information than the generation before.
3. There is not a single example of natural selection creating a new complex creature or even a new organ or functional part within a creature.
4. Natural selection cannot stop the tiny changes or downward degeneration of the genetic code within each subsequent generation of an organism.

Student Activity - Can Natural Selection Create a New Creature?

Make two black and white copies of a pencil drawing of some creature such as this frog (any creature will do).

+ + + + + + + +

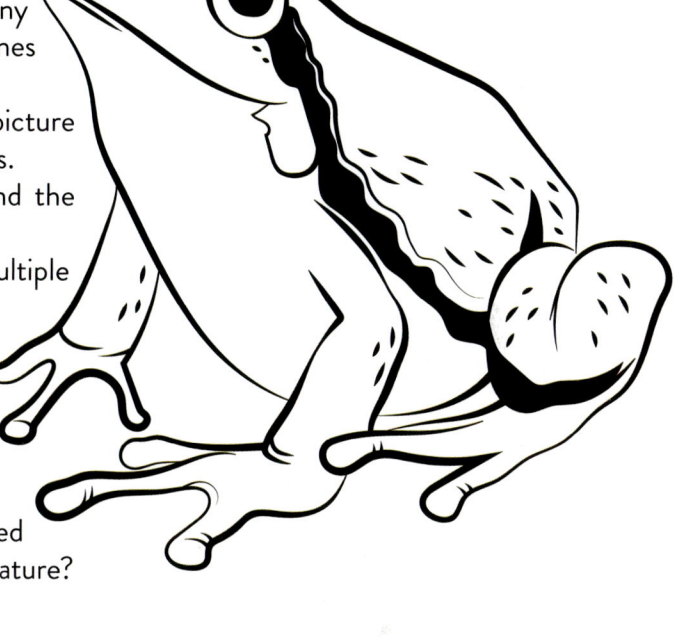

- Give the copy to student #1 and have them make 5 tiny changes with a pencil (small changes or modified lines only a few mm or fraction of an inch in size).
- Have student #2 erase the worst change and pass the picture to student #3 who will make 5 more random changes.
- Continue to have one student make 5 changes and the next erasing the worst one.
- After going around the entire classroom, maybe multiple times, compare the picture to the first picture.
- Explain to your students that their random changes are like mutations and natural selection is the process of erasing only the worst of the changes (it cannot correct small insignificant changes).
- Has this process of erasing the worst mistakes changed the creature into a better drawing, more advanced creature? Neither will natural selection!

Alternative Activity + + + + + + + + + +

- Take two copies of a picture of some animal and cut one copy into 50 pieces.
- Assemble the pieces correctly on a table.
- Have one student randomly move 5 pieces and a second student put one piece back to try and preserve the picture or create a new type of animal.
- Repeat the process.

Does the picture ever return to a normal animal or that of any other creature?

- Explain to your students that their random rearrangements are like mutations and natural selection is the process of correcting one of the changes (it corrects only the worse change).
- Has this process of erasing only the worst change preserved the creature or changed it into some more advanced creature? - Neither will natural selection!

a) Darwin's Finches

One of the classic "evidences" that one animal can change into a different kind of animal was an observation that Charles Darwin made in the Galapagos Islands in the 1830's. He noticed that common finches had one beak shape on the isolated islands but a completely different beak shape on the mainland. This and other observations of small variations within a given species led to his theory that different environments can cause small variations to have an advantage. In the case of finches, where long thin beaks allow birds to reach food better, this is an advantageous feature and Darwin taught that the finches "evolved" this type of beak. Whereas, if the birds live in a location where hard nuts need to be cracked open, thick heavy beaks have an advantage and the finches "evolved" this type of beak. Darwin extrapolated his observation that different finches have different beaks into the belief that one creature could evolve into a completely different kind of creature.

Bird beaks are made of hard keratin, the same material as our fingernails. The beaks grow continually throughout the life of a bird. England's Great Tits can change their beak shape effortlessly twice a year (thinner for summer insects and thicker for winter nuts). Oystercatchers were observed to change the shape of their beak—sharp or blunt— according to their food preference. It has been found that Hawaiian Honey Creepers will rapidly change their bill shape to accommodate differing flowers. This brings us to Darwin's Galapagos finches - used to convince generations of students that one form of life, by slow gradual changes over time, could change into a different form of life.[40,41]

Darwin used finch beaks to "prove" evolution. He saw minor changes within the beaks and then extrapolated. Darwin thought that given enough time, anything could turn into anything else. Darwin did not know about the incredible complexity of the cell or genetics; to his generation, the cell was just a blob of "protoplasm." Since then, scientists have been studying these Galapagos finches and have found that their bills change shape within the egg while the beak is forming - as the food source varies. The bird's bill shape is now thought to be governed by a protein called BMP4, which signals the molecule that controls the formation of bone. The amount of this protein present causes the finch embryo to produce either a thick stout beak or a narrow-pointed beak. These changes are not the result of mutations but are designed within the organism. The activation of the BMP4 protein allows the finch to adapt to the changing environment – depending on whether the environment is drier or wetter.[13] This is God's "programmed filling" in action. It allows the finch to fill changing environments.

In spite of this marvelous ability to vary for survival, the finch remains a finch – it simply has variations in the beak size. God created the bird kind with variation within each kind. With Darwin's Galapagos finches, and every other example of natural selection, we see variety programmed to adapt, but always within a given kind of creature.

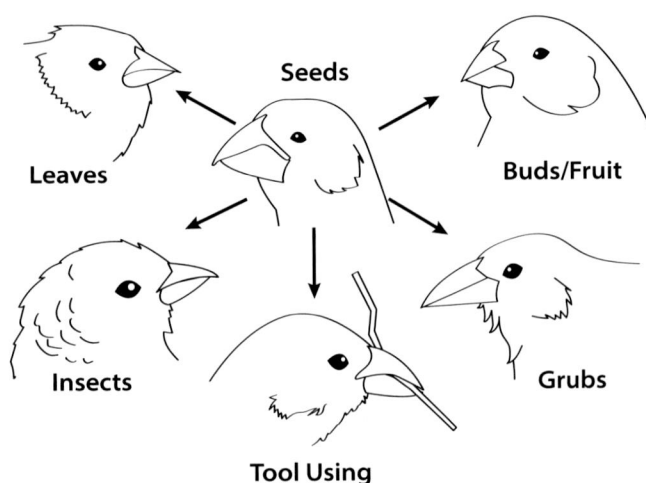

Darwin's Finches: Adaptive Radiation

Leaves, Seeds, Buds/Fruit, Insects, Tool Using, Grubs

+ + + + + +

+ + + + + +

b) Dog Variations

Students are often shown the amazing variety of dogs as an example of the capability of evolution to cause variations needed to change an animal's size, shape, and abilities. The characteristic of an animal's natural habitats, like vegetation and average temperature, can affect what type of dog thrives in a given location. For instance, in cold climates a long hair dog may do better, whereas in a hot climate a short hair dog may have an advantage. But does this show where the information came from which produced long hair and short hair dogs?

There are over 300 dog breeds in the world and the DNA of all domestic dogs has been shown to be traceable back to a single pair of gray wolves from the relatively recent past. There are several amazing implications of this discovery.[42,43]

First, it means that the unbelievable variety of modern dogs was present within the DNA code of the original gray wolf before the domestication of modern dogs took place. Thus, the huge variation in dogs; from Chihuahuas to Collies, Dalmatians to Dachshunds, or Poodles to Pointers, have nothing to do with "evolution." All of these varieties are simply a sorting of information already present within the "wolf kind." Part of the key to this amazing variety is that most dogs have 78 chromosomes, whereas there are only 48 chromosomes in chimpanzees. Having so many more chromosomes allows dogs to produce many variations within a short amount of time. **No new information is added; existing information is just rearranged** in countless ways to produce the breeds we see today.

Dog variation can be compared to a kaleidoscope - each turn produces a new pattern, yet the same number of beads remains within the kaleidoscope. Evolution requires new information to be added but that has never been observed - just variation within the dog kind.[14] Evolution must explain where this enormous pre-programmed information came from and it has utterly failed to do so.

Second, it illustrates the total unreliability of the evolutionary time frame associated with the appearance and modification of biological life. There is enormous disagreement and controversy between geneticists and anthropologists as to when the domestication of modern dogs happened, but all agree that modern dogs must have appeared less than 10,000 years ago. This is WAY too short of a time period for any significant evolutionary change to have occurred. Again, this confirms that dogs are a result of the sorting of pre-existing information.[15]

Student Activity - Explaining Dog Varieties?

Purchase kaleidoscopes for the students. I made my own from kits or you can look up how to construct low cost kaleidoscopes online.

Leave one kit not complete in order to show the "jewels" that are put in to make the designs. The same "jewels" are used but can make different designs.

+ + + + + + +

You can look up interesting stories behind the abilities of different breeds of dogs, including when and why different breeds were developed.

Each time a different breed of dog is mentioned (and what the specific breed can do), have the students turn the kaleidoscope to see another design. For instance:

- Rat Terries were bred during World War I to rid trenches of rats.
- Bull Dogs were bred in England to catch bulls, horses, cattle, boars and bear.

The designs all had the same "jewels," but were just rearranged.

c) Peppered Moths

Almost every biology textbook has the following example of natural selection in action. In England, before the industrial revolution, it was common to find the peppered moth in proportions of 95 percent light-colored to 5 percent dark-colored. This was believed to be caused by the majority of the trees in a certain area being light, so the light-colored moths were better camouflaged. Thus, fewer light-colored moths were eaten by predators. After the industrial revolution, the trees became primarily dark-colored (due to pollution) and the light-colored moths were now at a disadvantage to the predators. Thus, the peppered moth population shifted from light colored to 95 percent dark. Are they still peppered moths? Yes! They have not evolved into a different type of creature. This is an example of the powerful ability of an organism to adapt to their environment, not evolution in action.

But how far does this go to explain the development of completely different types of animals? We started with light and dark moths and we ended up with...Light and dark moths. Nothing new developed. The population merely shifted. Natural selection cannot add any new information to the genetic code of any organism. It is merely a process God built into nature to prevent the deterioration of life by allowing animals with advantages to survive in different environments. If that information is not already there, natural selection cannot make it appear! By the way, we now know that the moth pictures used in textbooks were glued to the tree to promote this evolutionary concept. Peppered Moths do not live on the trunks but in the canopy of the trees.[44]

There is not ONE example of natural selection producing a new type of animal, a new organ, or even a major permanent change in an existing organism. This is because "natural selection" is just that - selection. It can create nothing new. Organisms are preprogrammed to measure, analyze, and adapt to their environment. This is clearly intelligent engineering, the entire idea of "natural selection" is a misnomer. Natural selection cannot cause new useful information to be added to the DNA of an animal. Darwin was simply ignorant when presenting this mechanism as the explanation for life's development.[45]

Photo by Jerzy Strzelecki is licensed under CC BY 3.0

d) Drug Resistant Bacteria

A classic example students are given for believing in bacteria-to-human evolution is that bacteria can evolve to become resistant to drugs over time. This is a major concern for medical science as bacteria populations become increasingly resistant to known antibiotics, making them harder and harder to kill.

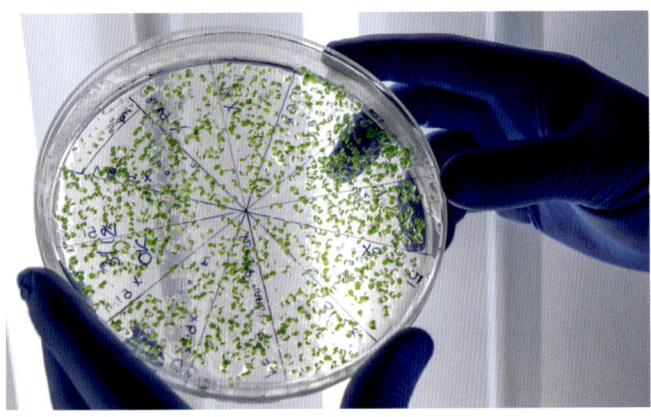

We actually have more bacteria in and on our bodies than we have cells. Furthermore, 99% of all bacteria appear to be non-harmful or even needed for life to exist. It is beneficial bacteria in our guts which help break down and process the food we eat. However, a small minority of bacterial types are harmful and even deadly. Once they start to multiply unchecked within our bodies, these harmful varieties can cause major problems.

This resistance is routinely credited to the bacteria evolving this ability to survive. There are two primary causes for this drug resistance and neither have anything to do with bacteria-to-human evolution.

First, as antibacterial agents kill 99.99% of a specific type of bacteria, the 0.01% which did not die already had the ability to resist the drugs. These drug resistant bacteria multiply to become an entire population of the same kind of bacteria, except the entire population is now resistant to the drugs. Nothing new developed, the bacteria resistant to the drugs just took over.

Secondly, bacteria are designed by their DNA programming to rapidly mutate, modify and adapt to changing environments. It has even been shown that bacteria can grab small sections of DNA from other bacteria and incorporate it into their "programming." Bacteria remain bacteria and do not evolve into some other type of creature.

1. Before and after any drug resistance developed, we started out with a certain strain of bacteria and ended up with the same strain – just with slightly modified abilities.
2. Bacteria are programmed to modify, rapidly adapt to mutations to become new strains, and even share DNA... but they remain bacteria.

Bacteria resistance testifies to design not evolutionary adaptation because bacteria are designed to adapt. This is creation, not evolution.[46]

KEY POINTS TO REMEMBER:

1. Examples of natural selection found in textbooks never add information to the DNA.
2. Natural selection allows a small population of animals to thrive only in a specific environment – it never creates a more advanced animal.
3. Darwin's finches are designed to vary in beak structure and can change back and forth as needed.
4. Wide variation in dogs is preprogrammed into their DNA. All dogs came from the information created within an original wolf.
5. Peppered moths remain moths. Both varieties were there originally and both varieties were present at the end.
6. Bacteria are designed to rapidly modify to survive in changing environments. They remain bacteria and do not evolve into some other type of creature.

B. "Beneficial" Mutations

Darwin's original theory of evolution included the idea that environmental changes could cause structural changes to occur in plants and animals. He also postulated that these acquired characteristics could be transmitted to offspring. In other words, a horse-like animal, by stretching its neck to reach the leaves in a tree, would be at an advantage if it had a longer neck. So, after a lifetime of using its body in this way, it might stretch its neck a little and would pass on this characteristic (which was acquired during its lifetime) to its offspring – eventually transforming into a giraffe. But any major changes like this are not written onto the DNA of the horse so they will NOT be transferred to the next generation. This original belief, known as Lamarckism, has been shown false and has been replaced by the belief that mutations are the driving force behind evolution.

Mutations are mistakes made during the transfer of information from the genes of one generation to the next (birth defects are examples of these). Believers in evolution postulate that if these mistakes are beneficial to the animal, it will give the mutated animal an advantage, and natural selection will then preferentially select these animals for survival. Although this belief seems logical, it does not fit reality.

Mutations are mistakes which have never produced an increase in complexity. Even examples of "beneficial" mutations, such as sickle cell anemia do not create new features. One hundred years of experimentation has shown that mutations cannot develop new organisms. This is because mutations never add an increased level of information. They are exactly analogous to random misspellings in a book. Therefore, this mechanism for evolution, even in combination with natural selection, fails to explain how new functional structures could arise.

For over 100 years, millions of fruit flies have been irradiated in laboratory experiments to observe the effect of mutations. The mutation rate has been increased by as much as 15,000 times[47]. The results of this experiment simulates millions of years of evolutionary progress. What has resulted are big-winged, small-winged, wrinkled-winged, and no-winged fruit flies; large-bodied, small bodied, and no-bodied fruit flies; red-eyed, speckled-eyed, leg-in-place-of-eye fruit flies; many bristled or no-bristled fruit flies; but mainly dead or sterile fruit flies. In conclusion, researchers began with fruit flies and end up with...well...Fruit flies - defective ones.

Furthermore, after several generations, even changes in the number of bristles on the irradiated fruit flies reverted back to the original number.[11] No new organ or useful functioning feature has ever developed.

The belief that mutations could slowly change an animal into some other animal is analogous to believing that an old vacuum tube black and white television could be changed into a 72 inch color flat screen by throwing random parts at it. The impacts will definitely produce changes but they certainly will not change the antique TV into a modern color flat screen. In the same way, mutations may produce changes, and it is remotely possible that some changes can find a function in a specific, limited environment, but they will not change an organism into some other type of organism. For that to happen, increasingly complex information would have to be added to the DNA of the creature. This simply is not going to happen as a result of random mutations.

Student Activity - Can Mutations Improve Things?

Select 10 students for this experiment. Have the first student draw a large picture of any animal of their choice on a full piece of paper without telling the next student what the animal is. Have the second student briefly look at the picture and then draw a picture of what they thought the animal was. Have the third student draw a copy of what they briefly saw when they looked at the second person's picture. Do not let them trace the previous picture but just draw their own version by memory. After just ten copies of the copy of the copy, have each student show their picture in order they were drawn. Have each student who drew a picture tell the class the name of the animal they thought they were supposed to draw.

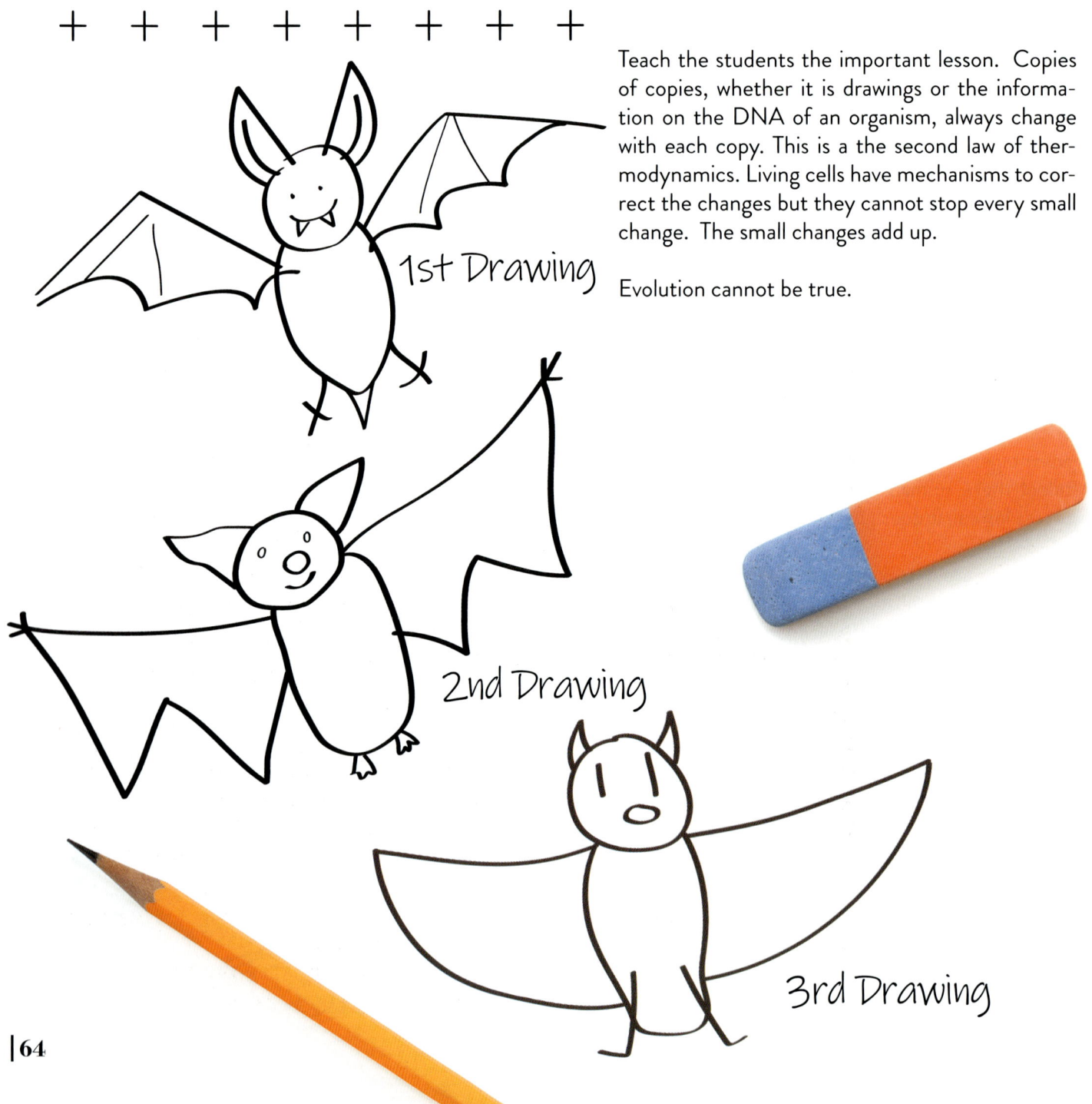

Teach the students the important lesson. Copies of copies, whether it is drawings or the information on the DNA of an organism, always change with each copy. This is a the second law of thermodynamics. Living cells have mechanisms to correct the changes but they cannot stop every small change. The small changes add up.

Evolution cannot be true.

Mutations as the driving force behind evolution have serious flaws which are seldom reported to students or to the general public. Let's take a closer look at a few textbook examples of mutations as a driving force behind evolution.

a) Sickle-cell Anemia

In almost every biology book, sickle-cell anemia is mentioned as a mutation which has a beneficial effect. This mutation only manifests itself as a deadly disease if both parents have the genetic defect. It results in deformed red blood cells incapable of carrying oxygen efficiently and early painful deaths for millions who have the genetic defect. The "benefit" of this horrible defect is that those suffering from it are more resistant to catching malaria. Thus, instead of dying an early

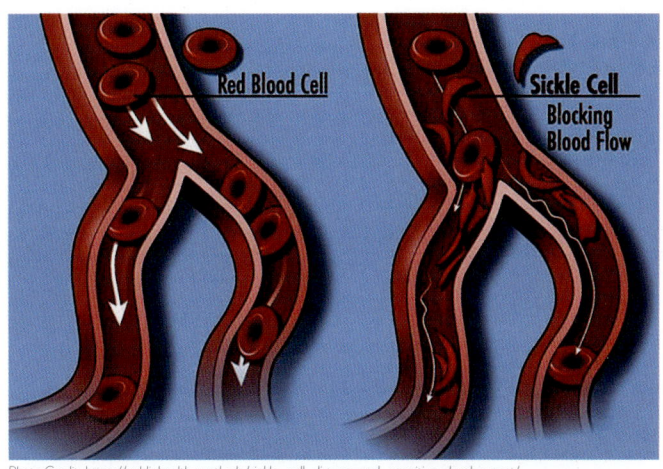

Photo Credit: https://publichealth.wustl.edu/sickle-cell-disease-and-cognitive-development/

death from malaria, they die early from another horrible, even less treatable disease – sickle-cell anemia.

In the words of an expert in this field of medicine, Dr. Chinmoy Biswas: **"This 'beneficial' mutation can in no way be construed as a gain of new genetic information that might support 'molecules to man' evolution.** *It causes deformity and malfunction of hemoglobin molecules, and its 'benefit' is the result of the interaction of two horrible diseases... Two copies of a sickle cell gene doom the bearer to painful disease and a shortened life span. But a single copy of the faulty gene can help the carrier survive malarial infection. This is because some red blood cells become faulty, but not enough to cause disease, and when these cells become infected with the malaria parasite they rupture, preventing the parasite from completing its life cycle. This reduces the number of malaria parasites in the blood and gives the carrier a slight advantage over non-carriers and they tend to be naturally selected to survive...* **In no instance is there an 'evolution' to a better and more advanced condition,** *with gain of new information and new capabilities."*[47]

The fact that such a debilitating disease is the best evolutionists can come up with as an example of a "beneficial" mutation is a stark testimony to how bankrupt the belief in evolution really is. This mutation does nothing to add new useful information to the genome and that is what is required for evolution.

b) Blind Cave Creatures

Many creatures which live in the total darkness of caves have a mutation which creates blindness. Blindness is actually an advantage in caves because a large percentage of the brain is dedicated to interpreting and understanding visual images. In total darkness this brain activity can be used for other functions - giving a creature an advantage. Blind fish, insects, and amphibians are common in caves. Evolutionists cite cave blindness as examples of "beneficial" mutations. They use such occurrences to justify their belief that other unknown "beneficial" mutations, over huge time periods, have

Photo by Arne Hodalič is licensed under CC BY-SA 3.0

transformed bacteria to people. Yet blindness causes no new information to be added to the genome of the creature and the loss of functional abilities (eyesight) is only "beneficial" in a very limited and specific location.

Student Activity - Blindness is Not Evolutionary Advancement

Share the following story with your students:
Imagine a movie theater filled with hundreds of people when the lights suddenly go out and someone screams, "FIRE," in a panic-stricken voice. Now suppose one person in the audience has been blind for his entire life because he was born with a genetic defect causing blindness. Of all the people now panicking in total darkness, which one will have an advantage?

Ask the following questions:
- Is the blind person more "evolutionarily" advanced?
- Does the blind person have a brain processing more or less information?
- Does this blindness show how bacteria could have changed into a person?
- If evolution is a fact of science, why are such misleading examples used (Such as blind cave creatures) to support the belief?

For years scientists believed that blind cave fish were the result of random mutations.

It has recently been discovered that cave fish were designed to go blind! As scientists were studying cave fish, they found that fish in the caves were almost identical to fish in the river outside the cave, except the cave fish had smaller or no eyes and were lighter in color. Actually, for those living in the cave this was an advantage, a highly developed visual system could use up to 15% more energy and soft eye tissue would be damaged if it bumped into the walls of a cave. Instead of eyesight, cave fish depend on their sense of smell and sensitivity to water pressure changes.

In reality, the blindness of cave fish is not the result of random mutations but of design. Conrad Waddington, a biologist, proposed that animals have a design mechanism which allows inactive genes to be "turned on" when environmental conditions change. Such a mechanism (a protein called HSP90) was found in cave fish. When the environment of a cave fish embryo experiences subtle factors such as low electricity in the water (caves have much lower electricity because the water has less salt), the HSP90 protein senses these outside conditions and turns off, causing the fish to lose its eyesight. Within a few generations, fish living in total darkness are born with scaled-over eyes or shallow unseeing sockets. This frees up their brain to process other senses more effectively.

Scientists were astounded to discover that when these same blind cave fish were introduced to water outside the cave, their offspring were born with fully functioning eyes within several generations! What scientists are confirming is that genetics are programmed to be flexible. But of course, God programmed within his creatures the ability to adapt to different environments.[48]

I'd like to end this section on "blindness" as an example of evolutionary advancement with a poignant but humorous story of one of many attempts to bring a more balanced education approach to origins into American schools. Over a decade ago, the State Board of Education in Texas was asked to include the evidence for creation and scientific problems with evolution into the public school science curriculum. Experts on creation were asked to testify and Don Patton, who has led multiple trips to Mt. Ararat in search of Noah's Ark, was one of many who came to testify. During questioning, he was challenged that blind cave salamanders showed a mutation which give these creatures an evolutionary advantage, proving that the mechanisms of evolution (mutations and natural selection) cause the advancement of life. Don's answer got right to the point as he responded in his classic Texas draw, "Well, I just don't see how goin' blind and livin' in a cave explains how bacteria turned into people. Now if that critter had evolved a head lamp, that would really have been something." Sadly, as with every other attempt to end evolutionary indoctrination in Western schools, the board (under immense pressure to conform to evolutionary beliefs) ultimately refused to allow students the freedom to see the evidence for creation.

Photo by Brian Baer and Neerja Hajelais licensed under CC BY-SA 3.0

c) Lenski's Bacteria

Professor Richard Lenski, from the University of Michigan, has been tracking changes in 12 identical starting populations of E. coli bacteria for 32 years. In that time there have been over 70,000 generations. One of the widely reported and publicized results of this experiment was the 2008 change in one of the strains which mutated to become a variation of bacteria which used citrate as its food in aerobic (oxygen rich) environments. Some news reports publicized this with the sensational headlines such as, "Bacteria evolve ability to eat plastic." Citrate ($C_6H_8O_7$) is not technically a plastic but is a short-chain carbon-based molecule. It is part of the ager solution used to grow bacteria but not normally the primary source of nutrition for the bacteria. The fact that a mutation allowed the bacteria to start using this molecule as a source of nutrition is widely reported to students as proof that evolution can create new abilities within organisms.

What is NOT told to students:

E. coli are already programmed to metabolize (eat) citrate in an anaerobic (oxygen poor) atmosphere and metabolize other nutritional sources in aerobic (oxygen rich) atmospheres. The bacteria have always had (i.e. they were created with) an internal "switch" which turns on the ability to process one or the other type of food sources depending on the amount of oxygen present. The nutrient agar that the bacteria are grown in contains citrate molecules making it easier to process food sources but, the citrate is normally ignored. In this long-term experiment, many mistakes (mutations) happened over time as generation after generation of bacteria made copies of themselves. After about 30,000 generations the bacteria's internal "switch" broke in one strain of the bacteria - causing this strain of bacteria to prefer citrate in an oxygen rich atmosphere rather than only using it in an oxygen deprived atmosphere. This gave them an advantage to grow rapidly in the artificial laboratory environment.

NOTE: These mutant bacteria would never survive in competition with normal bacteria in a normal environment. They ONLY have an advantage with the artificial laboratory food source because the other bacteria in their vicinity were not using that food source.

What actually happened was a loss of information and capability. Bacteria which originally had the ability to use two different types of food in two different oxygen environments now only had the ability to use a single type of food. It is a loss of information and ability, not an evolutionary advancement.[49]

Once again, for this type of observation to be promoted as proof that bacteria could evolve into people, shows a complete void of actual scientific evidence evolutionists have to support their belief system. No mutation ever increases the total information content in DNA.

Student Activity - Random Changes never improve a book

Share with the students that DNA is exactly like a language. The chemicals that make up the molecule are like letters on the page of a book arranged in a specific order. It carries meaning and has a purpose. Languages never write themselves – they require intelligence.

+ + + + + + + +

Write this paragraph (or something like it) in large letters so it completely fills a piece of paper:

> Life is so complex that it has to have had a Creator. This Creator holds us accountable for knowing he exists by observing creation (what He has made.) Languages never write themselves and random mistakes never improve books. Mutations are mistakes that never increase the complexity of life.

Write in light pencil so the letters can be written over during the activity.

On another piece of paper put all the letters in the alphabet so that they fill the page.

Pass the two papers around and have each student in turn:
1. Close their eyes and randomly place one fingertip on each page.
2. Replace the closest letter in the paragraph with the randomly selected letter from the other page.
3. Pass the two pages on to the next student to do the same thing.

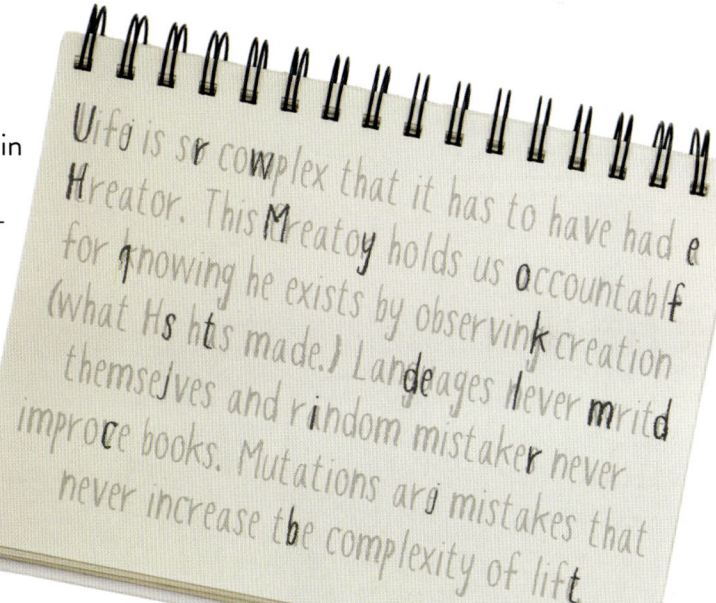

Once the page is done, show it to the students and try to read the new message.

Ask your students:

- Have random changes caused the information content to increase?
- Do mutations add useful increased information?
- Based on observations like this, could evolution be possible?

KEY POINTS TO REMEMBER: + + + + + + +

1. Mutations always destroy information.
2. Sickle-cell anemia is a disease which kills people but gives a resistance to malaria. It is not an evolutionary advancement.
3. Blindness of cave creatures is a loss of ability.
4. After 70,000 generations of bacteria, the bacteria are still bacteria, even with the mutation.

C. Common Evolution "Proofs"

A wide variety of examples are given in biology and other textbooks to promote the belief in evolution as a fact of science to students. Many of these examples are just variation within a kind of creature via information already found within their DNA, or examples of natural selection allowing variations of creatures to survive in a specific environment. We have already examined a few common examples of "evolution." These are called "evolution," and when scientists state that evolution is a "fact of science," they are referring to genetic variation or natural selection. But these types of examples are explained as well (or better) by creation and in no way prove that completely new types of creatures can form or that bacteria have turned into people. There are a few other "proofs" of evolution that tend to show up in textbooks.

a) Vestigial Organs

Vestigial features are those parts of an organism thought to be useless leftovers from its ancestor, as the creature has evolved to a new way of life. Our tailbone, appendix, and tonsils were all commonly taught to be such features. The idea of vestigial features has been used as purported evidence for evolution since 1859 when Darwin first proposed that such features were evidence for the descent of one organism from one completely different. The logical consequence of this alleged transformation is that the "new" creature would be left with some features which are no longer needed in its new environmental niche.

Belief in evolution demands that we believe that each type of animal on earth is a result of its descent from some previous life form. If this were the case, almost every creature should have many leftover features which are no longer needed. Yet, the more we learn about biology, the more we discover that every part of an organism serves some useful function. For example, the appendix is often said to be a useless leftover from our supposed ape ancestry. We now know that the appendix serves as a type of lymphatic tissue in the first few months of life to fight disease. It also stores good bacteria so that when stomach flu or other diseases clear out our digestive tract, our appendix can "reboot" the good bacteria into our small intestines. The appendix serves multiple functions and is never listed in anatomy books as a vestigial organ. Yet it is still claimed to be so by evolution believers and biology textbooks.

The acceptance of the idea that some parts of the human body are useless leftovers has led to many tragic consequences. Based on the misguided concept that the human colon was a vestige of the past, Sir William Land and dozens of other surgeons stripped the colons from thousands of patients in order to "cure" a variety of symptoms. Many died. Tonsils were considered useless and as late as the 1960's many people routinely had them removed. This practice was again fueled by the mistaken belief that the tonsils were a useless leftover feature from our past. It is now known that they serve an important disease fighting function and should not normally be removed.

Have you ever wondered why the end tip of the human spinal column is called the tailbone? The name stuck because evolutionists believe that this is a leftover part of human evolution when we supposedly had tails and this leftover part is useless. Really? Do you like to sit, stand, and go to the bathroom? The "tailbone" or more accurately called the coccyx is enormously important, for it is the place to which tendons, ligaments, and muscles attach. The 12 muscles that are anchored here help you in such functions as standing erect, supporting pelvic organs, and assisting in bowel movements. This "organ" is neither leftover nor useless. We need to be calling it our coccyx because we never had tails. "Ouch, I just fell on my coccyx!"

The human body is designed for maximum versatility -- it is far more versatile than the body of any other creature. What other animal can perform the range of movement required for activities as diverse as ice skating, pearl diving, pole vaulting, snow-boarding and gymnastics? This range of movement would be impossible without the coccyx!

In summary, evolution predicts that there should be leftover features as one organism slowly turned into another. Based on this theory, over 150 human features were at one time listed as useless. Dr. Jerry Bergman has written a book which conclusively shows that every vestigial organ, which was used to justify evolution, has now been proven wrong![51] Creation predicts that although some designed parts may have degenerated, resulting in a loss of original function, every part of an organism was designed to serve some useful purpose. If evolution were really true, there should be thousands of useless and no longer needed parts and chemicals in every organism - but none are found.[51] Assuming parts are useless has slowed scientific advancement and cost many lives. As we learn more about the biology of living organisms, including ourselves, it is readily apparent that creation better fits the evidence.

+ + + + + + +

CERVICAL SPINE

THORACIC DEPARTMENT

LUMBAR SPINE

SACRUM

COCCYX

KEY POINTS TO REMEMBER:

1. If evolution were true, there should be thousands of now useless features in every kind of organism as it slowly evolved upward. Over 100 human organs were listed as vestigial (or useless leftovers) in the late 1800s.
2. Not a single one remains today – the appendix, coccyx, and tonsils all have useful functions.
3. Death and agony have resulted from this misguided evolutionary belief.

b) Junk DNA

DNA is the coded language which allows every organism to build the parts needed for every cell to function and every animal to reproduce copies of itself. The parts of cells in every organism are made of many different complex chemicals and the information to make these chemicals is found on the DNA molecule. Yet less than 10% of the information on the DNA molecule is used to make these complex proteins and biochemicals. What is all the other coding, or language, on the DNA used for? For many decades this information was referred to as "junk DNA." The faulty assumptions of evolution slowed the discovery of many DNA functions. If we evolved from previous life forms, there should be vast stretches of coded information on the DNA molecule with no current functional purpose.

The only view shown to students is the idea that all this coded information is the result of random mutational changes over time. Thus, as a bacterium turns into a sponge, it no longer needs the "bacteria" programming as it now depends on its "sponge" programming. As the sponge turns into the fish it no longer needs its "sponge" or "bacteria" programming. All this "old" information on the DNA would become useless or "junk" DNA. Thus, if evolution were true, there would be no reason to look for any current purpose for all that leftover, useless junk programming.

Researchers were in for a surprise. Essentially, every part of the DNA molecule has some functional purpose. Some sections are repeated for redundancy. Other sections are for error checking. Some sections allow for storage of backup information. Some areas form breaks in information just like paragraph and sentence breaks. Certain sections of DNA form structural areas allowing it to efficiently split apart during cell division.

Scientists now realize the idea of "junk DNA" is not true and have thrown it into the junk heap of false science. The belief in evolution, resulting in the belief in "junk" DNA, significantly slowed scientific advancement in the area of genetics.[52,53]

KEY POINTS TO REMEMBER:

1. If the DNA of one creature slowly transformed into another there should be lots of now useless, unneeded programming.
2. This misguided belief stopped researchers from even looking for the function of DNA sequences of unknown purposes for many years.
3. Even though mistakes are constantly building up within the DNA coding, essentially all DNA sequences are increasingly being acknowledged to exist because they serve some useful function and purpose.

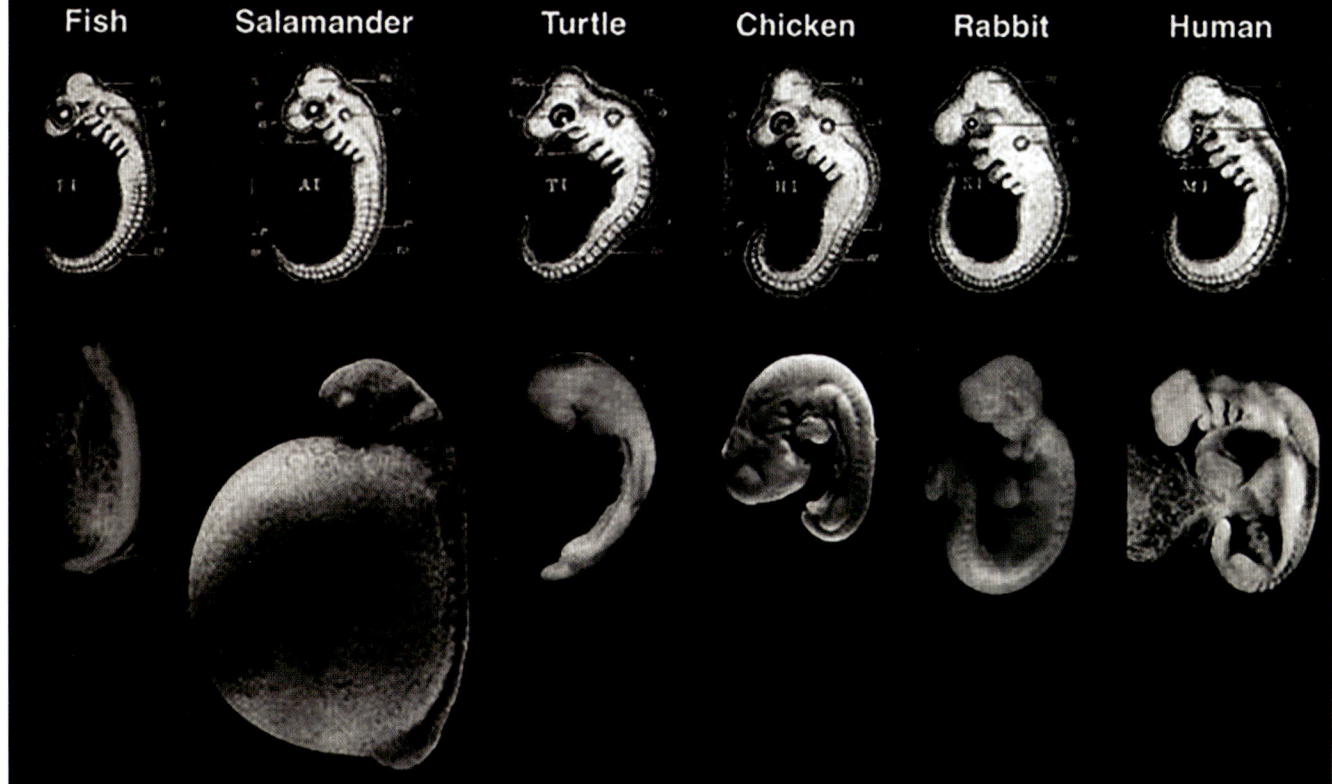

©Springer-Verlag GmbH, Heidelberg, Germany.
Top row: Haeckel's fake drawings of several different embryos, showing incredible similarity in their early 'tailbud' stage. Bottom Row: actual photographs by Michael Richardson of the very different animals at this stage of development.

c) Embryo Similarity

A common consequence of the belief in evolution can be found in most high school biology books that contain a section on comparative embryology. This is the idea that humans and animals have a common ancestor because their embryos have a similar physical appearance. This concept was popularized in 1866 by Ernst Haeckel as he traveled throughout Europe showing the similarity of different animal embryos. As early as 1894 he was reprimanded by his own university for using fraudulent drawings.[54]

Yet unbelievably, Haeckel's embryo drawings from 1866 are still in the textbooks, some 140 years later. Are these drawings correctly drawn? NO! Embryologists have known since 1894 that Haeckel's evidences were faked. In the illustration below (published in 1997), Haeckel's drawings are shown on the top row and the actual embryos are shown on the bottom row. Notice that they do not match! Amazingly, his teachings remain in textbooks to this day. Since the 1950's, it has been proven that a woman's fertilized egg is a complete human being. Only time and nutrition are required for it to grow larger. From the moment of conception, a pregnant woman's body is two bodies, not one. Women do not give birth to fish, salamanders, turtles, chickens, or rabbits. The second body from the moment of conception can never be anything but a human.

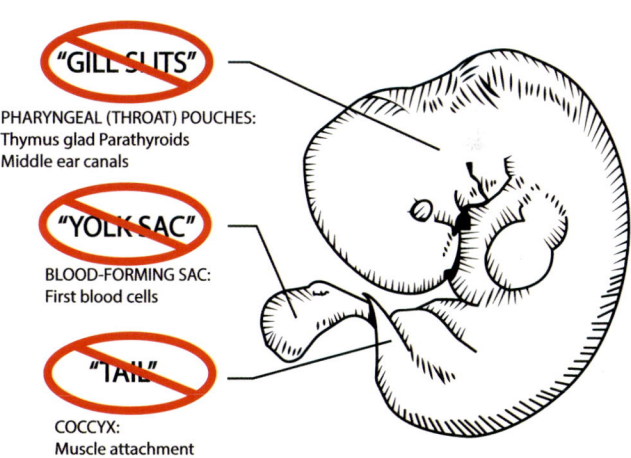

Examples of developing baby parts have also been said to be evidence of evolution. Students are told that developing babies go through an evolutionary history with "gill slits" and "yolk sacs" like a fish or a reptile. In a growing baby, the "gill slits" and "yolk sac" serve completely different functions from those of animals from which they are supposed to be descended. Gill slits form gills in fish. In humans, they are merely folds forming various glands and facial features.

The same is true of the "yolk sac."

The yolk sac contains food for a reptile or a bird; while for a human, it has a radically different function. In a growing baby, the heart and circulatory system develop before the bones (which will ultimately be the baby's blood source). A baby's heart actually starts to beat 18 days from conception. Yet, the developing baby may have a different blood type than its mother, so it cannot use her blood. With no bone marrow to make its own blood, how can the baby continue to develop? The simplest engineering answer would be to provide a temporary alternative supply. The yolk sac does exactly that, then it disappears! [55]

KEY POINTS TO REMEMBER:

1. Haeckel faked the drawings of embryo evolution.
2. Human embryos do not go through stages of evolution inside the womb. Various features turn into the parts of a human body, not parts of some other animal.
3. Haeckel's drawings have been unscientific from the beginning, over 140 years ago.

d) Fossil Evolution - Whales, Horses, and others

Over the years, many creatures have been promoted in textbooks as showing slow gradual development over time as they "evolved" into modern animals. Lining up various shapes of similar creatures, (or lining up fossil remains of different similar creatures) in order to "prove" evolution has a glaring logical error. Suppose I was to line up a bicycle, motorcycle, and automobile. By lining up these three very different forms of transportation have I proven one turned into the other all by itself? Of course not. Yet this is exactly what textbooks do – they show students sequences of fossils or variations of similar animals and lead students to believe that this proves that evolution has happened. It is evidence of a common designer and the vast variability of information within the DNA code of a given creature – NOT evidence for evolution of one type of creature into another. Let's take a closer look at some examples still found in textbooks:

i) Horse Evolution

In the early 1800's, Europe had a strongly Christian culture. Very few doubted the existence of a Creator or questioned the Bible as an authoritative source of truth. Today, Europe, Australia, New Zealand (and increasingly so America) operate largely as atheistic cultures, where the Bible is considered irrelevant.

First to fall was the timeline of the Bible, as Charles Lyell, a lawyer-turned-geologist, proposed slow and gradual processes over eons of time as an alternative to the worldwide flood. This idea replaced the acknowledgment of a global flood as the explanation for the geological features of our planet. Following close on his heels was Charles Darwin, proposing that all of biology could be explained by slow gradual changes over eons of time as an explanation for the biological diversity of our planet. This direct attack on the authority of God's Word did not achieve immediate acceptance because it failed to explain the development of highly complex organisms. Nor did it explain how enormously complicated and interdependent organs such as the eyeball could have developed gradually. To sway the masses, evolution proponents needed a visual, easily understood example of "evolution in action" which could be grasped by both the highly educated student and the man on the street. They needed an icon around which to rally. They found their publicity gold mine in the horse.

Up until the early 1900's, horses were THE mode of transportation used by individuals and industry alike. Everyone was familiar with the common horse. In 1876, T.H. Huxley (commonly known as Darwin's Bulldog because of his avid promotion of evolution to the scientific community) visited an arrangement of horse fossils collected and arranged by Yale University paleontologist Othniel Marsh. Huxley immediately recognized the potential of this sequence for popularizing evolution. Here at last was an icon of evolution with which everyone could connect - a sequence of creatures showing a small, four-toed, "pre-horse" creature with short teeth changing into a modern horse. This was "evolution in action," dug right out of the rock record, mounted and displayed for all to see. The famous horse sequence soon filled textbooks and museums around the world. Imagine the impact on the belief and thought processes of a generation who were totally dependent upon the horse for their very economic survival.

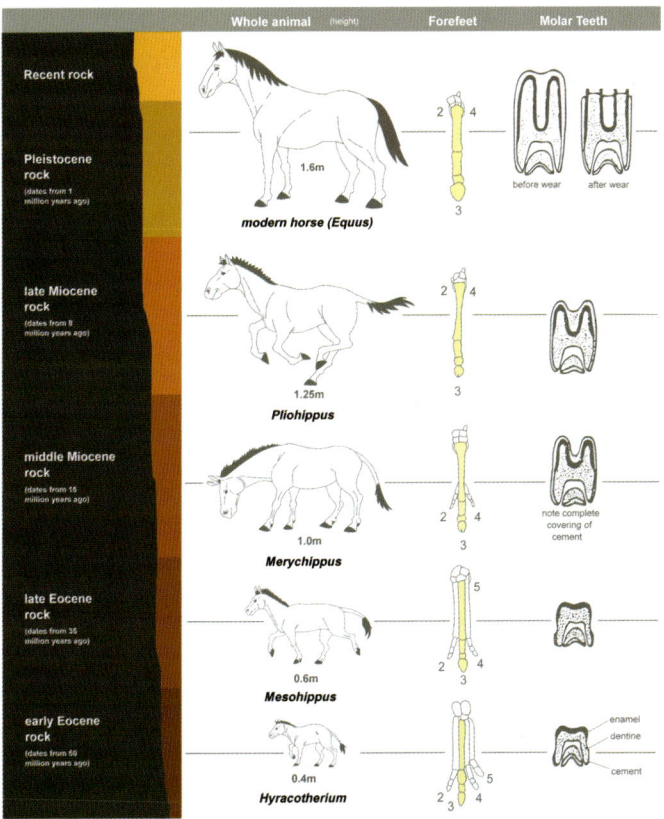

Original: Mcy jerry Vector: SchlurcherBotPixelsquid, CC BY-SA 3.0 via Wikimedia Commons

The 'horse evolution' story begins with a dog-sized, four-toed "dawn horse" or "eohippus" moving on to the larger three-toed Mesohippus, then to a still larger three-toed Merychippus and finally to our modern horse with one toe, Equus.

It should be noted that "dawn horse" was discovered by Richard Owen. He thought it was similar to a modern-day hyrax or rock badger, thus naming it Hyracotherium (it was later called eohippus or "dawn horse.") So, the first "horse" in the evolutionary sequence was named after a hyrax or rock badger because that is what it looked like – a small badger, not a horse!

It is only now, over a hundred years after the damage had been done, that evolutionists sheepishly admit that the horse sequence was largely fantasy. For instance, Dr. Niles Eldredge, former Curator of the American Museum of Natural History, has stated, "There have been an awful lot of stories, some more imaginative than others, about what the nature of that history [of life] really is. The most famous example, still on exhibit downstairs [in the American Museum] is the exhibit on horse evolution prepared perhaps 50 years ago. Now I think that is lamentable, particularly when the people who proposed those kinds of stories may themselves be aware of the speculative nature of some of that stuff."[56]

The smallest ancestors to the modern horse have been found in the same rock layers as essentially modern horses. Today's horses can be small like "Thumbelina," a miniature brown mare standing at 17 ½ inches tall to a Belgian draft horse standing at 6 feet 7 ½ inches. So, the size of the horses does not prove evolution, it just shows a variety of different size horses. Other "evidence" for horse evolution – changes in the number of toes and teeth design have similar problems. Three-toed horses are still alive today, and the shape of horse teeth can vary in both modern horses and fossils of extinct variations. Yet all these variations are still horses. The "evolutionary sequence of horses" reveal variation within the horse kind not molecules-to-man evolution.

KEY POINTS TO REMEMBER:

1. The "horse series" is not showing molecules-to-man evolution, but a wide variety within the horse kind.
2. Both horse fossils and modern horses vary in size, toe number, and teeth configuration.

ii) Whale Evolution

Another "proof" of evolution which is often shown in textbooks are the fossilized bones of whales which have a tiny little extra bone in the pelvis area – this is supposedly the remnant of their ancient legs. Scientists have found that the tiny pelvis bone functions as

Whales are perhaps the most remarkable group of animals to go back to sea, and their evolutionary journey is now quite well understood due to a variety of fossils found mostly in Pakistan. The animals represented here are not to scale and don't represent a direct line of descent, but rather plausible models for how this amazing transition happened.

Indohyus major
48-40 mya - Pakistan

Pakicetus inachus
52 - 48 mya - Pakistan

Ambulocetus natans
47 - 41 mya - Pakistan

Rodhocetus kasrani
48 - 40 mya - Pakistan

Protocetus atavus
45 mya - Egypt

JULIO LACERDA

- A whale's pelvis is shaped completely different from land animals. Whales do not have legs.
- All of its muscles and attachment points on a whale are different from land animals - allowing an up and down tail movement.
- Its entire breathing apparatus would have required the migration of holes in its skull to move from the front to the top of its head.
- Its body would need to develop the ability to produce insulating blubber (different from land animal's fat).
- Whales have a different temperature regulation process than land animals.
- Whales have a special breathing and metabolism system.
- Whales use a radically different blood/oxygen exchange system (counter current).
- Whales have a special design for mating and giving birth underwater.
- Dozens of other radical changes from land animals to whale could be listed.

If any item on this list were just partially evolved as a land animal changed into a whale it would doom the transforming creature to a watery death. How would the creature survive when it is ½ land animal and ½ whale? How would a whale survive if there were no fins yet? How would they survive if the blowhole was in front? Also, in the fossil record whales are whales. No true transitional forms have been found.[58]

part of the reproduction apparatus, not as ancient legs. Evolutionists also claim that whales can breathe because their blowhole migrated from the front of the skull to the top of the skull. No whale fossils have been found showing this blowhole migration (it is just assumed to have happened).

Darwin speculated that some sort of bear-like creature ventured back into the ocean to become today's modern whales. Since his time, others have speculated that everything from cows to wolf-like creatures to hippopotamus turned into whales. All of this is fantasy, not science. For a land animal to turn into a whale would have required thousands of useful and specific changes as it somehow transformed from a land-dwelling creature into an ocean dwelling creature.

This lack of transitional forms was one of the major problems which bothered Charles Darwin about his theory from the very beginning. He stated that, *"... the number of intermediate varieties, which have formerly existed, should be truly enormous. Why then is not every geological formation and every stratum full of such intermediate links? Geology assuredly does not reveal any such finely graduated organic chain; and this, perhaps, is the most obvious and serious objection which can be urged against the theory. The explanation lies, as I believe, in the extreme imperfection of the geological record."*[19a]

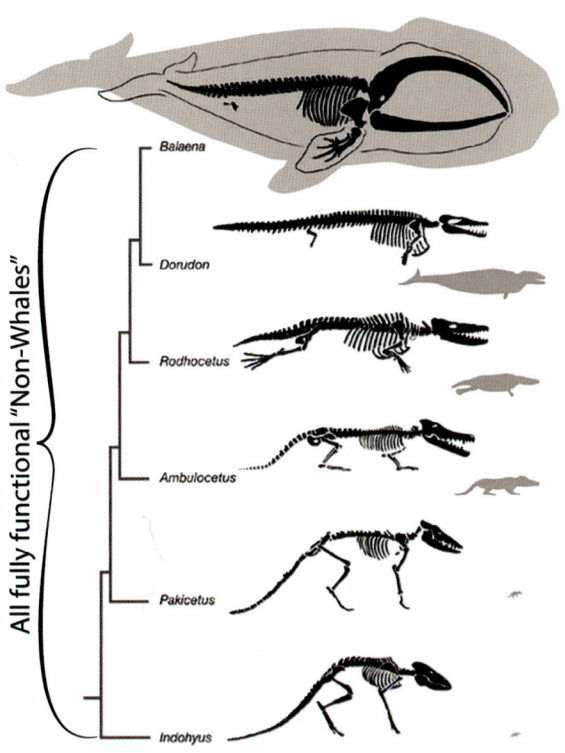

We now have millions of fossils, and the situation today is no better. Here are a few quotes from paleontologists who are honest enough to acknowledge what the fossil record actually shows:

- Dr. Colin Patterson, senior paleontologist at the prestigious British Museum of Natural History, and the author of the book, Evolution, made the following written comment when questioned why he did not include any illustrations of transitional forms in his book, *"...if I knew of any, I certainly would have included them..."* [59]
- Stephen J. Gould was a leading proponent of evolution at Harvard University for many years and made this statement: *"All paleontologists know that the fossil record contains precious little in the way of intermediate forms; transitions between major groups are characteristically abrupt."* [60]
- Paleontologist David Kitts admitted that, *"Despite the bright promise that paleontology provides of 'seeing' evolution, it has presented some nasty difficulties for evolutionists, the most notorious of which is the presence of 'gaps' in the fossil record."* [61]
- Evolutionist Mark Ridley wrote: *"...no real evolutionist, whether gradualist or punctuationist, uses the fossil record as evidence in favor of the theory of evolution as opposed to special creation..."* [62]
- Steven Stanley writes: *"The known fossil record fails to document a single example of phyletic evolution accomplishing a major morphologic transition "* – i.e. no significant change in body structure has ever been found in fossils. [63]

So why do textbooks still tell us that fossils support evolution? Evolutionists believe that the evolution happened so fast that it was not captured in the rock layers. This is given the technical sounding name of "punctuated equilibrium." Animals supposedly stayed the same for huge period of "equilibrium," then there was a "punctuated" rapid change which happened so fast that the transitional forms were not captured in the rock layers. Do we see punctuated evolution happening today? Students are now told evolution is happening too slowly to be observed. So, which is it? Evolution is the only area of "science" where the lack of evidence supporting it (no transitional forms) is presented as evidence in support of it!

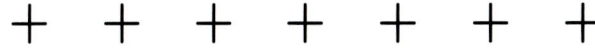

KEY POINTS TO REMEMBER:

1. Whale fossils look just like modern whales; no evolution found.
2. All major transitional forms are systematically missing. It is assumed in the past that evolution happened TOO FAST to capture these transitions. Yet today, evolution is supposedly happening TOO SLOWLY to see major changes happen.
3. Evolution is not science, but rather an illogical belief.

D. Human Evolution

The lineage of transitions from some sort of ape-like creature into human beings is shown in textbooks as if it is a fact of science. This icon of evolution has been shown to students for over 50 years, and convinced many that God has nothing to do with the origin of humanity.

The textbook portrayal of half-humans scraping together a bare existence of berries and prehistoric animal meat is so common that most people believe that there is total agreement in the scientific community over this concept of our past. However, this idea is based on conjecture rather than fact. There is actually considerable disagreement exists as to its validity. Listed below is a very brief summary of just a few of the key ape-to-man links which have been used to "prove" human evolution.

Neanderthal (*Homo neanderthalis*)

After Darwin's theory of evolution was published in 1859, the search for man's ape-like ancestor began in earnest. Throughout Europe, many apparently human skeletons were found which had thicker than normal bones, eyebrow ridges and slopped forehead. These skeletons were reconstructed with a hunched over, ape-like appearance and were presented to the public as "ape-man" links in spite of the fact that many experts of the day disagreed with this conclusion. If fully clothed and placed in a modern city, it is unlikely these people would even be noticed. The differences in bone structure are easily attributed to pathological diseases or minor genetic variations.

Monkey Business

A reasonable explanation for finding remains of Neanderthals in the caves of Europe is the Flood of Noah causing a worldwide ice age immediately following the Flood. All of the volcanism and land movement during the Flood would have poured trillions of BTU's into the ocean water resulting in greatly increased evaporation for centuries following the Flood. There is much evidence for a wetter environment worldwide during the ice age, and warmer oceans would have resulted in this increased evaporation. Further north in Europe, this rain came down as enormous amounts of snow causing the one and only ice age which occurred upon the Earth. The ice age would have lasted for centuries - until the oceans of the world cooled down, resulting in less evaporation and less cloud cover. This ice age likely forced those moving into Europe (Neanderthals and others) to take shelter in caves as sheets of ice increased over time.

Neanderthal bones and archaeological sites reveal that they used tools, hunted in coordinated groups, used musical instruments, sewed with needle and thread, buried their dead, painted beautiful pictures on cave walls, and spoke with complex language. In other words, they were fully human! Neanderthal Man is now classified by almost all scientists as fully human and is merely a minor variation of the human family.[64]

"Java Man," "Peking Man," "The Hobbit" and others (<u>Homo erectus</u>)

Peter Maas, CC BY-SA 3.0 via Wikimedia Commons

Hundreds of skulls and fragments have been found with smaller brains and a more sloped forehead than modern man. These are classified in a general category called homo erectus, and routinely presented to students as advanced apes on the way to becoming humans. These bones are often found with tools or habitation structures and were obviously intelligent.

In 1892, Eugene Dubois found a thick boned skull cap in the same general vicinity as a human thigh bone. This is still presented as evidence of an early transformation from ape to man. However, many years after this supposed creature was widely accepted as an "ape-man" link, Dubois admitted that the skull cap and leg bone were separated by 46 feet in a gravel deposit. They could easily have been from 2 different individuals – an ape and a human. These details are omitted when Java man is discussed.[65]

In the 1920's, a group of researchers found a large number of ape-like skull fragments in a cave near Peking, China in the direct vicinity of fire pits and tools. Although all of the original skulls disappeared during WWII, it is still assumed that since the skulls and tools were found together, this was an ape-man link. However, students are seldom presented with a more plausible explanation. Monkey meat is very tough, but monkey brains are still considered a delicacy in that part of the world. Since only the skull fragments were found, it is quite likely that Peking man was man's meal ... not man's ancestor.[66]

The Hobbit is the nickname of Homo floresiensis, found on the island of Flores, Indonesia in 2004. They were small-brained, four foot tall, perfectly proportioned humans. They used tools and would had to have constructed sailing vessels to reach the island they were found on. Every indication is that they were fully human, and many experts now agree with this interpretation of their bones.

Many other Homo erectus bones are presented in textbooks as proof that man has evolved from apes. Almost every year, National Geographic, Time Magazine, and science periodicals feature some new discovery fitting into this category. In actuality, every find can be shown to be either a misinterpretation of the evidence or a small deformed, inbred human population struggling to survive during the ice age period after the Flood of Noah. They were not apes turning into humans, but degenerating human beings. The best explanation for the many homo erectus bones is the degeneration after generation of humans who were isolated, starving, and inbreeding after spreading out across the world from Babel. It is a known biological fact that small populations that experience inbreeding and limited food sources result in smaller brains and distorted body structures. This is so well known in biology it has been given a name – insular dwarfism. The human like "apeman" bones found by scientists are not apes in the process of turning into humans, but isolated and degenerated humans. Other even more apelike bones are simply that – variations of now extinct apes. The best book documenting these facts has been written my Christopher Rupe and Dr. John Sanford called Contested Bones.[67]

Lucy (Australopithicus)
The most common creature presented as the missing link between apes and humans is known as "Lucy," and given the scientific name, "Australopithecus." This very ape-like creature supposedly preceded Homo erectus in the evolutionary progression from ape to man. The most famous example was found in the early 1970's by Donald Johanson and brought him instant fame. The 40% complete set of bones was missing most of the skull, hands, and feet. Lucy was found in a layer of volcanic rock later found to contain footprints identical to human footprints. So, Lucy was given human feet in subsequent museum displays. Since that time, more Australopithecus remains have been dug up, and they have had chimp feet.[67b] It is thus still debated whether this creature walked upright in a human manner. Many respected evolutionists reject the claim that "Lucy" was an ape-to-man link. For instance, British anatomist Lord Solly Zuckerman conducted an extensive examination of a wide variety of Australopithecus fossils, concluding that they were not upright walkers. He and a team of scientists spent 15 years studying the anatomical features of humans, monkeys, apes, and Australopithecus fossils before coming to this conclusion. If "Lucy" did not walk upright; the obvious conclusion is that Lucy was an ape. Further studies by Dr. Charles Oxnard also came to the conclusion that Australopithecus were not intermediates between man and ape, but simply an extinct ape.[68] There are similar problems with every fossil link between humans and apes.

Almost every year the media and popular magazines publicize the latest ape to man link. There is almost always a big announcement of the "major" new find, but with time, it quietly fades to obscurity. Not only is there considerable disagreement between evolutionary researchers, but the evidence is sparse, fragmentary, and open to other interpretations.

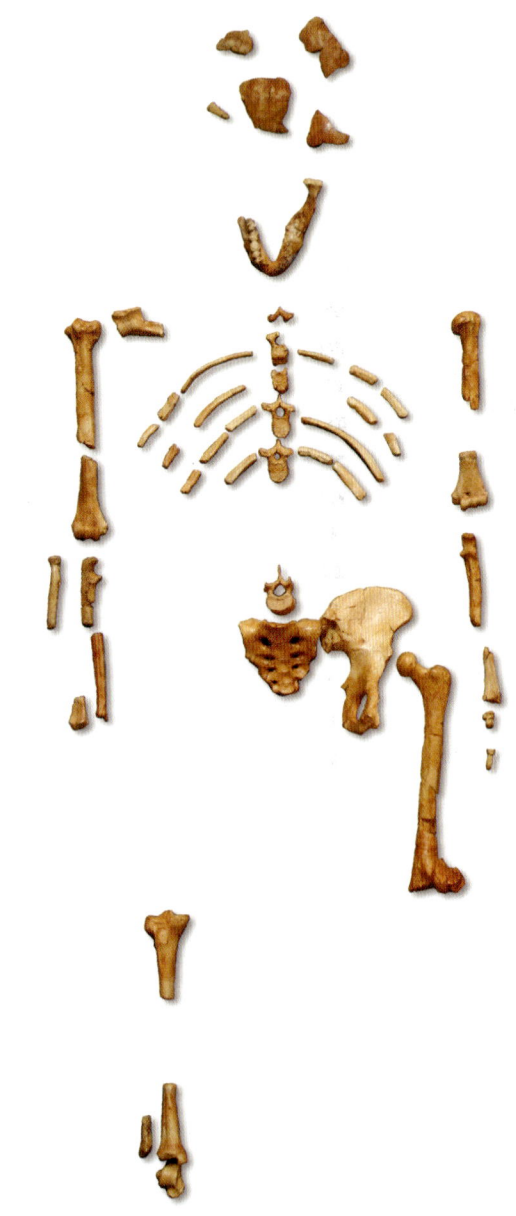

120, CC BY-SA 2.5 via Wikimedia Commons

KEY POINTS TO REMEMBER:
1. Neanderthal was fully man, not part ape/part human.
2. Homo erectus such as Java man, Peking man and hobbit were a blend of fossils, apes, or degenerated people - not part ape/part human.
3. Lucy and other Australopithecus were simply extinct varieties of apes.

Here is a partial list:

Neanderthal man	Fully human	Australopithecus	Ape variety
Cro-magon man	Fully human	Lucy	Ape variety
Homo engaster	Degenerated human	Homo rudolfensis	Ape variety
Heidleberg man	Degenerated human	Ardi	Ape variety
The Hobbit	Degenerated human	Sediba	Ape variety
Ardi	Orangutan variety	Ida	Lemur variety

Teacher Demonstration - The Nose Test

The human nasal bones protrude from the skull, while ape nasal bones are flat: a pair of glasses sit nicely on our noses but will slide right off an ape's face.

Engage your students to test this idea to determine if skulls presented as human-like are really human or just variations of apes and monkeys.

+ + + + + + + +

You will need a model of an ape's head, a model of a human head, and a pair of glasses. You can possibly find an ape and human mask from Halloween and use these as your models. Or you can sculpt prototype skulls with wire mesh, coated with paper and paste. Show how glasses fit on a human head but slide right off an ape or monkey's head.

Once the students see how the glasses sit on humans and not on an ape, show pictures of other supposed ape to human links and ask, "Can he wear glasses?" Then ask what is he - an ape or man?

Even this one small difference makes it really clear which is which!

> Can they both wear glasses?
> Does this make them an ape or a man?

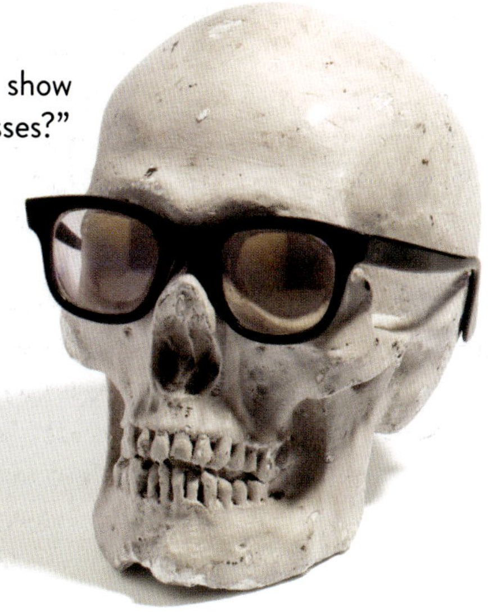

SECTION V - PROBLEMS WITH THE COSMIC EVOLUTION

The idea of a big bang creating all of the matter, energy, time, space, stars, galaxies, planets, and ultimate life in the universe is repeated endlessly throughout the media, movies, and school curriculum. Therefore, it is simply accepted as being true by most people. The Big Bang Theory says there was absolutely nothing, then all the energy and matter in the universe appeared as a single point which explosively expanded. This matter turned into gases which spread out (along with space itself) throughout the universe. These clouds of gas pulled themselves together to form stars and galaxies. The stars created other elements, which condensed to form planets, and life eventually appeared on these planets.

The reason cosmic evolution teaches that nothing turned into everything is because, if evolutionists start with something, they have to acknowledge a Creator. Since cosmic evolution must assume that everything made itself (or that time, space, and matter have existed eternally), it starts by believing that nothing turned into everything during the Big Bang. But how does this belief hold up to scientific scrutiny?

Questions:
- Where did the energy come from?
- How did it get so compressed to form a single point? (The Natural Gas Law shows that gases always spread out if not confined.
- What caused the explosion?
- Do explosions make order and design? (Explosions just make big messes.)
- What was around before the Big Bang? Evolutionists say there was nothing, because if they start with "something" they have to explain where the "something" came from.
- Do I get something from nothing?
- Does this equation sound logical: Nothing + Nobody = Everything

It takes more faith to believe in the Big Bang Theory than to believe that God recently created the universe as stated in Genesis. God was there at the beginning of time and He told us what He did: "In the beginning, God created the heavens and the earth." Genesis 1:1. There is nothing unscientific or illogical about this belief. Furthermore, God tells us exactly how and in what order He created things. Creation involves faith but so does cosmic evolution, but evolution violates known and testable laws of science whereas creation does not.

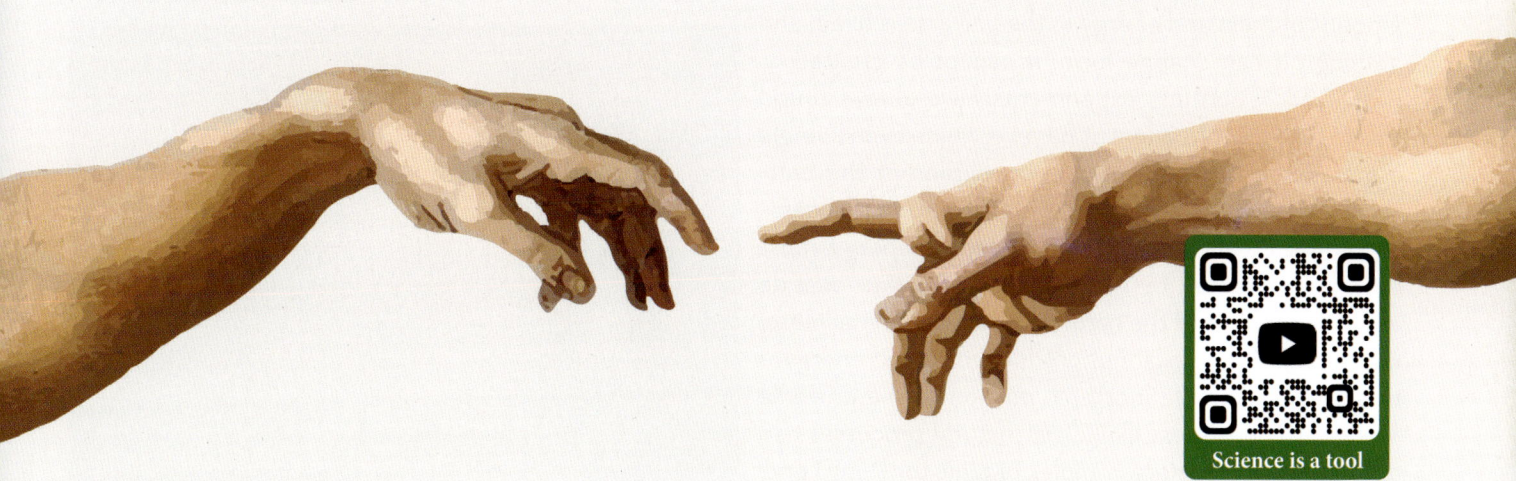

Science is a tool

Combining evolution with Biblical teachings makes a mockery of both. Compare the Big Bang with the Days of Creation:

Evolution	Bible
Nothing became everything	Everything exists because it was created
The universe had evolved over billions of years	An all-powerful God created recently
Stars formed by themselves before the Earth	Earth was made first, then stars later
Sun formed before the Earth	Sun formed on fourth day
Plants evolved after the sun	Plants were created before the sun
Information (DNA) and life created itself	Information requires intelligence
Dinosaurs died 60 million years before man appeared	Man and dinosaurs lived at the same time

Evolution and the Bible cannot both be true. Evolution and creation are two completely different worldviews. Let's take a closer look at some of the enormous unsolved problems with cosmic evolution (i.e., the Big Bang), and what actual scientific observations reveal.

A) Can Nothing Turn Into Everything?

The First Law of Thermodynamics states that the total amount of mass and energy in the universe is fixed, and although it can change form, it cannot be created or destroyed. Countless experiments have always found this to be true. No scientist has ever made even a single atom appear from nothing. A conclusion of this conservation of mass and energy is that natural processes cannot create energy. Consequently, energy must have been created by some agency or power outside and independent of the natural universe. Furthermore, if natural processes cannot produce mass and energy – the relatively simple inorganic portion of the universe – then it is even less likely that natural processes can explain the much more complex organic (or living) portion of the universe.

The only logical alternative to the First Law of Thermodynamics is to assume that the universe itself is eternal and has always existed. Yet, this contradicts the Second Law of Thermodynamics that says the universe is "winding down." Examples of this are everywhere:

- Shuffling cards – they never return to perfect organization.
- Perpetual motion machines – They are impossible because usable energy is always lost.
- Springs and swings – They eventually come to a stop after being stretched or swung.
- Heat – Hot objects eventually cool to the surrounding temperature. Everything cools down, yet stars are still hot.
- Stars – They are used up, run down, and explode but we do not see them reforming.
- Life – It wears out and dies, but we never see new life forms appearing from nothing.
- Genetic mistakes – Always increase with every subsequent generation.

Countless experiments have also always found the Second Law of Thermodynamics to be true – the universe is winding down and wearing out, not improving in complexity and order. If the universe is winding down, who wound it up?[69]

KEY POINTS TO REMEMBER:

1. Cosmic evolution and the Big Bang directly contradict what the Bible tells us.
2. The 1st Law of Thermodynamics is the most tested law of science and states that nothing cannot become something.
3. The 2nd Law of Thermodynamics shows that the universe is winding down, not increasing in complexity all by itself.
4. Testable, observable science shows that the universe could not be eternal (The 2nd law) and couldn't have made itself (the 1st law). So evolutionist have to chose which law of science to ignore.

B. Can The Big Bang Explosion Create Order?

Explosions never create order; they always destroy order. If you detonate a bomb in a pile of building materials, does it create a building? Obviously not. Explosions never create order. Yet the universe is finely tuned to perfection and far more complex than any building. If the strength of atomic constants, the charge of electrons, the strength of gravity, the total mass of the universe, and dozens of other constants of science were changed even slightly, the universe could not exist. The universe gives every indication of having been perfectly tuned and designed in its current state such that life could exist. It is statistically impossible for all of these constants of physics to have happened by chance.

This is why it is increasingly popular for cosmologists (those studying the formation of the universe) to talk in terms of infinite and multiple universes. Infinity is a mathematical concept that allows the belief that time is limitless, and everything, no matter how improbable, will eventually happen. There is absolutely no evidence for multiple or parallel or an infinite number of universes, yet this concept is repeatedly showing up in popular movies such as Spider-man, Avengers, Star Trek, Fringe, and many others.[70] Just as in these movies, this idea is science fiction and not scientific reality.

The Big Bang: Speculation and Guesswork

KEY POINTS TO REMEMBER:

1. Explosions do not create order, but our universe is guided by extremely ordered laws and constants of science. Therefore, the Big Bang explanation of its existence cannot be true.
2. Infinite universes or "multi-verses" are fanciful beliefs in order to escape the reality that the universe is too perfect to have created itself.
3. Since there is no way to test the idea of another universe existing, this is a religious belief, not science.

C. Can The Universe Explosively Expand?

The Big Bang is based on faith, not science.[71] Here are a few examples:

- One of the critical mathematical requirements of the Big Bang model is that the universe expanded rapidly (greater than the speed of light) early in its formation, and then slowed down at precisely the correct moment in order to allow stars to form. This is called the "inflation period." No one has ever observed anything moving faster than the speed of light – **it is just assumed to have happened because if it did not happen the Big Bang would not work.**

- No one can explain why the expansion would slow down at precisely the correct moment for stars to form – **it is just assumed to have happened, because if it did not happen the Big Bang would not work.**

- The transformation of energy to matter in particle accelerators always results in equal parts of matter and antimatter, yet our universe consists almost entirely of matter. No one has ever explained where all the antimatter disappeared to – **it is just assumed to have happened, because if it did not happen the Big Bang would not work.**

- The condensing of gas to form stars has never been seen to occur. Gas clouds in space are always seen to be expanding, not contracting. To solve this problem, it is assumed that over 90% of the matter and energy in the universe is "dark matter" that can neither be seen or tested. It is this unseen matter that is compressing gas clouds to create stars. No one has ever explained where all this dark matter and energy came from or proved that it even exists – **it is just assumed to be there, because if it does not exist the Big Bang would not work.**

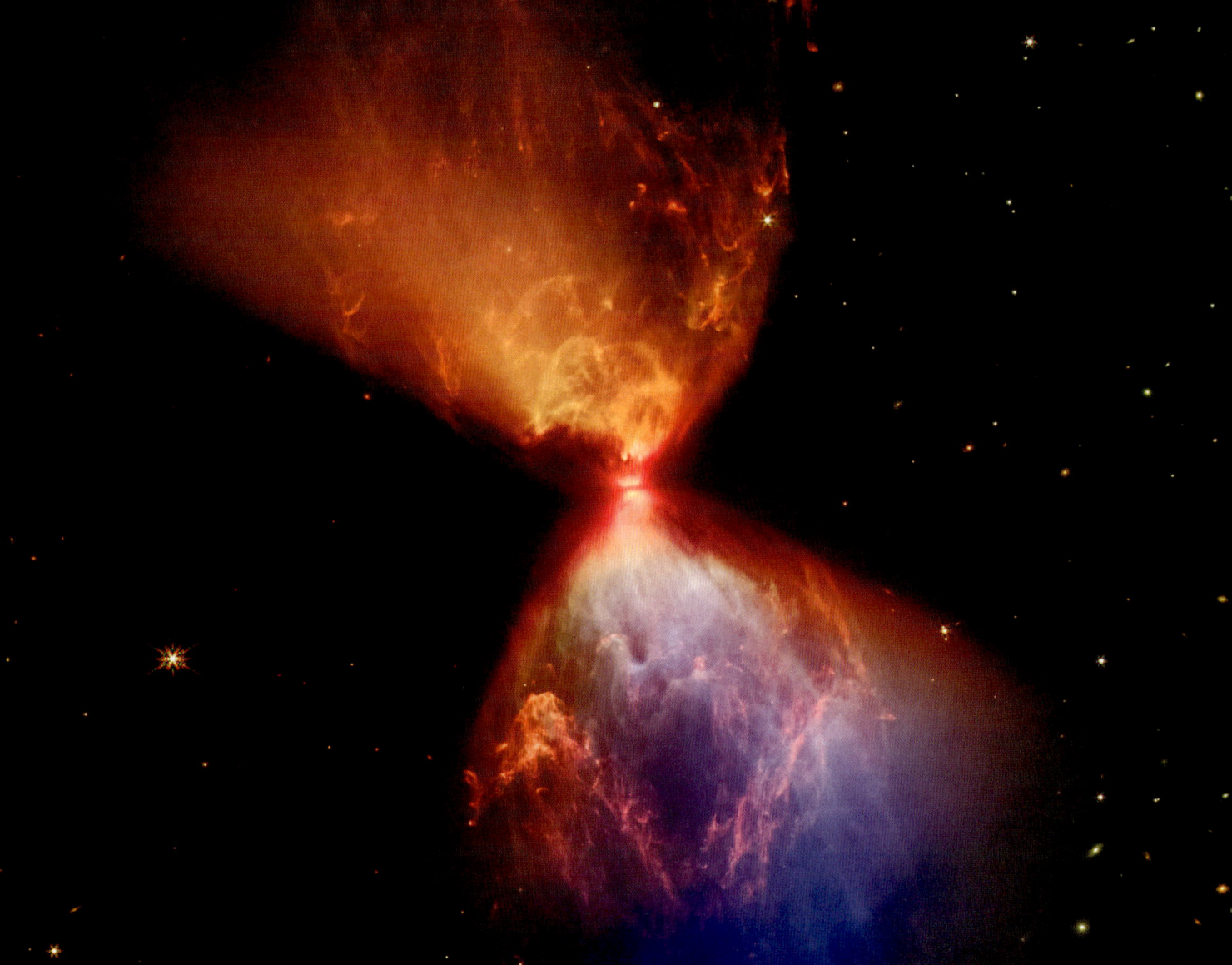

Photo: NASA, ESA, CSA, and STScI, J. DePasquale (STScI)

D. Can The Universe Explosively Expand?

One of the most tested laws of science is the Natural Gas Law that states, left to itself, gas molecules will always spread out from high pressure to low pressure. There is nothing in space to contain gas molecules, and in the vacuum of space, gas is billions of times more spread out than here on earth.

Gravity weakens with the square of the distance between particles. Thus, atoms twice as far apart have only one fourth the attractive force of gravity. In other words, the forces of the Natural Gas Law far exceed the attraction of gravity for condensing gas together. Stars have never been seen to form, and the laws of science tell us they will not form by themselves. [72]

KEY POINTS TO REMEMBER: + + + + + + +

1. The Big Bang has many problems which are solved by assuming unobserved and impossible things have happened in the past.
2. The Natural Gas Law shows that stars could not form themselves.
3. No new star has ever been seen to appear.

Student Activity - Can Gas Compress to Become a Star?

Place an empty balloon on your desk and pass around a filled up balloon. Have the students gently squeeze the balloon.

+ + + + + + + +

Ask the students the following questions:
- When you squeezed the balloon did it seem to push back against your hands?
- If you separate your hands in open air and move them together, does the air push back?
- Is the "air pressure" higher inside the balloon or within the open room?

Now ask the critical question.
- Will air molecules, all by themselves, ever flow into the balloon to fill it up?
- If we open the balloon, will air flow from higher pressure (inside the balloon) to lower pressure (outside the balloon) OR will the open balloon get bigger and bigger as more and more air packs itself into the opened balloon?

> Demonstrate this concept by making a small hole in the balloon with a pin. It will explode as all the air rushes out to lower pressure! Explain to students that it is impossible for gas molecules in space to pack together to form a star, just like it is impossible for a balloon to fill itself up.

Teacher Demonstration- The Natural Gas Law and Star Formation

Explain to the class that those who believe in evolution believe that gravity pulls hydrogen gas together in space to form new stars. You are going to test this idea with an experiment.

Show a can of compressed air (or hair spray) to the class. Explain that there is high pressure gas inside the can and lower pressure gas outside the can

- The Natural Gas Law states that gas ALWAYS moves from high pressure to lower pressure.
- For a star to "make itself", gas must move from lower pressure (the vacuum of outer space) and pack itself billions of times closer together into a higher and higher pressure area to form a star.
- That would be like opening the valve in the can and having all the air in the room compress itself inside of the can!

If you believe in the Big Bang, then gases must have compressed to form stars...that formed elements...that formed planets...that formed complex chemicals... that came alive... that formed us. If evolution is true, we are just stardust and stars must have made themselves.

Now open the valve to see if air molecules from the room will compress themselves into the can - like a star making itself. Tell the students to hold their breath because all the air in the room may be sucked into the can and they may die.

Push the valve on the can and ask the students which way the gas went: *Out of the can into the room or did the lower pressure room air pack itself into the higher pressure can?*

> This happens every time, because it is a law of science. Gas could never condense to form stars and no scientist has ever observed a new star form. God made the laws of science so we will KNOW He exists. Certain things could never make themselves:
> **Matter and energy, Stars, and Life**

SECTION VI - THE AGE OF CREATION

A. Dating Methods

It is hard to open a newspaper, book, magazine, or watch a movie without finding some implication to the Earth being billions of years old. Given the overwhelming barrage of these statements, it is understandable why so many people have trouble considering the possibility that the Earth might be only thousands of years old. Yet, there is an overwhelming amount of data that indicates that the earth is indeed much younger than the billions of years claimed by evolutionists.

Despite what people have been led to believe, there is no dating method which gives an absolute date for the formation of the earth. All dating methods are based on non-provable assumptions about some event in the distant past. Furthermore, there is a strong bias to reject any dating method which does not allow the time evolutionists require. To understand the validity of any date, your students need to understand how all dating methods work. The following illustration should help:

Time = Amount/Rate

Amount = Measured - Initial - Contamination
Rate = Average Rate Over the Whole Time

Suppose you were up at 6:00 a.m. and happened to see a friend who lives in a nearby town. You observe that he is walking along at 2 miles an hour and you know that he lives 16 miles away. You can easily use the formula at the top of the illustration to calculate that your friend left home 8 hours earlier. You have just performed a dating method of how long your friend has been on the road. However, something doesn't make sense. Why would your friend be up all night walking? Although you used the correct formula, your assumptions may not have been correct. Perhaps your friend stayed with someone close to you and woke up minutes before for a morning stroll. In this case, you have used the 'wrong initial amount' in your calculation. Perhaps he took a shortcut which cut 4 miles off his walk. In this case there was 'contamination' of the total amount. Perhaps since you last saw your friend, he has taken up marathon running and averaged 8 miles an hour (only having slowed down just before you saw him). In this case, you have used the wrong 'Average Rate.' The point is, wrong assumptions lead to wrong answers.

In all dating methods, the initial amount is an assumption, the estimate of contamination is an assumption, and the overall rate is an assumption. The only things which can be known for sure are the present amount and the present rate. Unless you estimate the initial amount correctly, the average rate correctly, and the amount of contamination correctly, your answer will be wrong. Furthermore, depending on your assumptions, it could be very, very wrong.

There are actually very few dating methods which seem to indicate that the earth is extremely old. On the other hand, nine out of ten dating methods indicate that the earth is quite young. If the earth is relatively young, creation becomes the only logical alternative. This is the primary reason that only those methods which seem to indicate the Earth's very old age are shown to students.

KEY POINTS TO REMEMBER:

1. All dating methods are based on assumptions.
2. Most dating methods indicate a recent creation

There are two common classifications of dating methods which illegitimately seem to indicate old ages:

1. Radiocarbon dating is used for organic materials which still contain carbon from once living things.
2. Radiometric dating is used for igneous rocks. Examples are uranium-lead, potassium-argon, rubidium-strontium, and many others.

B. Radiocarbon Dating

In order to understand why carbon-14 (^{14}C) dating gives erroneously ancient dates, we must understand how this dating method works and the assumptions used for interpreting the results.

When one of the sun's high energy particles hits a nitrogen nucleus, neutrons are released which react with ^{14}N to form ^{14}C. This ^{14}C atom rapidly reacts with oxygen to form a $^{14}CO_2$ molecule. Plants take in the $^{14}CO_2$, animals eat the plants, and a uniform dispersion of radioactive carbon throughout the biosphere (all living organisms) results. As long as an organism is alive, it continues to take in both ^{14}C and ^{12}C in this ratio. However, after it dies the radioactive carbon begins to decay and the ratio of ^{14}C to ^{12}C decreases. It takes 5,730 years for 1/2 of the original ^{14}C to decay. Thus, if something has a ^{14}C to ^{12}C ratio of 1:2 instead of the current 1:1 it is assumed to be 5,730 years old. A ratio of 1:4 would be interpreted as 11,460 years old, 1:8 would be 17,190 years old, etc. Highly sensitive mass spectrometer methods have improved the ability to measure the amount of ^{14}C so that 0.001% of modern amounts can be measured. This equates to 18 half-lives. If the ^{14}C generation level was always constant, this would correspond to an organism about 100,000 years old. This is the actual; detectable limit.

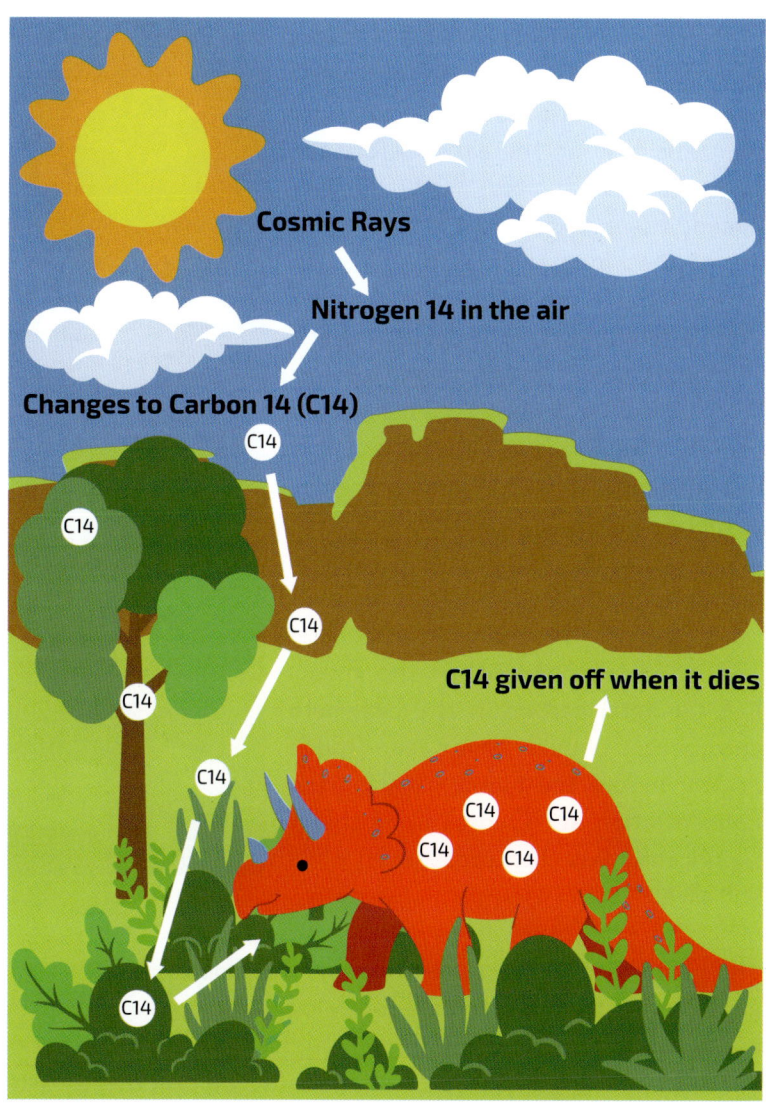

The decay rate of ^{14}C is such that after 250,000 years, starting with today's concentration, every molecule of ^{14}C should be gone. Astonishingly, since 2005, samples from "300 million years old" carbon-containing coal and other fossils has been shown to contain measurable levels of 14C, averaging 0.3% modern of ^{14}C levels. This means almost every known fossil and organic sample is not millions of years old, but extremely young. The presence of this radiocarbon is conclusive evidence for a recent worldwide Flood, which buried these organic materials about 4500 years ago.

Original amount of carbon-14		
$\frac{1}{2}$ remaining	$\frac{1}{2}$ decayed	5230 Yrs
$\frac{1}{4}$	$\frac{3}{4}$ decayed	11,400 Yrs
$\frac{1}{8}$	$\frac{7}{8}$ decayed	17,190 Yrs
$\frac{1}{16}$	$\frac{15}{16}$ decayed	22,920 Yrs

Why are ^{14}C levels significantly lower for plants and animals alive 5000 years ago? There was less ^{14}C being produced in the atmosphere because a stronger magnetic field on the earth before the flood, resulted in less ^{14}C being produced. There is significant evidence that during the Flood of Noah the strength of the earth's magnetic field dropped dramatically, and it is still decreasing at a significant rate.[74a] This resulted in much less ^{14}C in all organisms alive before and shortly after the Flood than there is in modern organisms. During the global Flood, enormous quantities of plants and animals were buried. Also, massive amounts of carbon dioxide precipitated out of the oceans as calcium carbonate. The world readjusted after the Flood, but the total carbon in the biosphere and oceans would have been much greater before the Flood - making the concentration of ^{14}C in all living things significantly less than today, even while still alive.

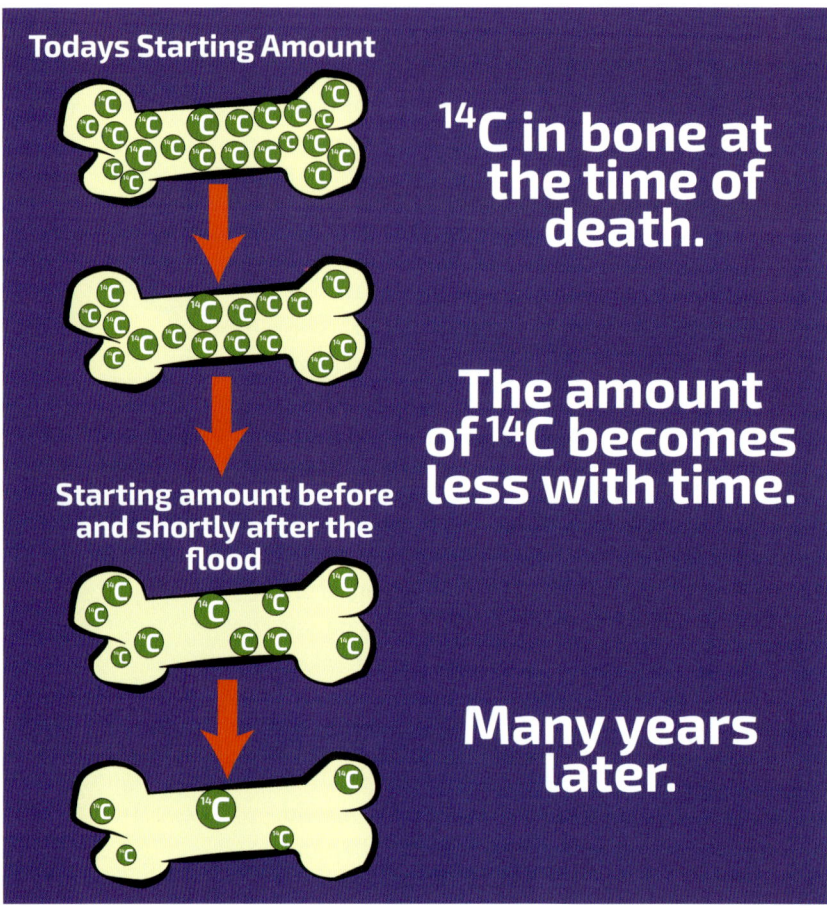

The Flood (and for centuries following the Flood when the atmosphere was rapidly re-adjusting) caused organisms living at that time to have much lower amounts of ^{14}C than today's amount. The assumption that the ^{14}C to ^{12}C ratio has been the same throughout history is an incorrect assumption. The Flood occurred less than 5,000 years ago, making any plant or animal dated before, shortly after 5,000 years ago, completely inaccurate. Fossils of plants and animals from this time period will actually date much older than they really were.

The real challenge to evolution believers is to explain why there is ANY radiocarbon left in coal, wood, gas, dinosaur bones, shells, etc.. which are supposedly millions of years old. Carbon-14 dating actually gives extremely compelling evidence that life on Earth is quite young.

KEY POINTS TO REMEMBER:

1. Radiocarbon dating of artifacts buried near the time of the Flood of Noah gives erroneously wrong dates because the concentration of radiocarbon shifted after the Flood.
2. The fact that there is radiocarbon left in old artifacts such as coal or dinosaur bones is strong evidence that they have <u>not</u> been buried for millions of years.

Teacher Demonstration - How Long Does Radiocarbon Last?

Give each student a piece of paper and a pair of scissors or demonstrate this principle from the front of the class.

Tell the students the paper represents all the radiocarbon in a living organism.
Once the organism dies, half of the radiocarbon disappears in about 5,000 years.

Cut the paper in half and have them count the number of years that passed (~ 6,000)

Cut what is left in half again (~ 11,000 years)

Cut it in half a third time (~ 17,000 years)

Keep doing this and have them count the years that passed.

Set one piece aside after 7 cuts

After about 17 cuts, this is the limit of our detection equipment. This would represent about 100,000 years. Set this tiny spec aside.

After about 36 cuts not a single molecule of radiocarbon (or the paper!) Will be left. Compare how much ^{14}C is left the (7 cut pieces) with how much should be left if fossils are as old as we are told (less than the tiny spec.)

HERE ARE TWO INCREDIBLY IMPORTANT POINTS TO SHARE WITH THE STUDENTS:

1. It is an assumption that things in the past started with the same amount of radiocarbon as things alive today. They did not. There are valid reasons to believe that before and after the Flood, organisms still alive had only 1/10th of the radiocarbon as today. Thus, something "dated as 20,000 years old" is really only 5,000 years old.

2. If the fossil has carbon, such as coal or soft tissue in dinosaur bone, is dated millions of year old, then not one single molecule of ^{14}C should be left! The amount that is actually left indicate a recent burial and destroy the evidence for an ancient Earth and evolution.

C. Radiometric Dating Used for Igneous Rocks

There are a wide variety of isotopes of the elements in the periodic table which are not stable because they contain extra neutrons in the nucleus. Thus, these "parent" elements fly apart to become other "daughter" elements over time. The rate at which this happens can be accurately measured. If this rate is very slow, and lots of the resulting daughter element is found in a rock layer, the rock is assumed to be extremely old. Think of this as sand falling through an hourglass. The more sand at the bottom, the more time passed since the glass was turned over.

As radioactive elements decay, one of the most common resulting fragments of the nucleus to decay is called an alpha particle. These are pieces of the nucleus of an atom which consists of two protons and two neutrons. These particles rapidly grab two electrons to become a very stable helium molecule. Since the original location of the radioactive elements is often within cooled rock layers (such as granite), the helium which forms from this radioactive decay is also trapped inside these rock layers. Many of these granite layers are assumed to have been formed millions of years ago - so the radioactive decay would have happened shortly after the formation and cooling of the granite. Thus, if the rock layers were millions of years old, any helium found within these rock layers would have also been trapped for millions of years.

Let's think about helium balloons. They float because helium is an energetic molecule which is lighter than air due to the low density (weight) of the gas. Helium balloons do not float indefinitely because the helium molecule is so small that it slips right through the wall of the balloon and escapes to an area where the concentration of helium is lower. This is why helium filled latex balloons sink within hours and even aluminum coated mylar balloons stop floating within weeks. Helium is such a small energetic molecule that it can escape at a measurable rate through solid stainless steel (i.e. it has a high permeation rate). This allows helium to be routinely used to measure the leakage rate through other materials. If you found a helium balloon floating on the ceiling of an ancient Egyptian tomb, you would know that someone had recently been in that room. If your guide told you that no one else had been inside the tomb in the last 4,000 years, you would know that they were either ignorant or lying. This brings us back to the helium found in granite across our planet.

Isotope		Half-life of parent (years)
Parent	Daughter	
Carbon 14	Nitrogen 14	5,730
Potassium 40	Argon 40	1.3 billion
Rubidium 87	Strontium 87	47 billion
Uranium 238	Lead 206	4.5 billion
Uranium 235	Lead 207	710 million

The amount of radiometric decay that has happened within any granite containing zircon crystals can be calculated by knowing the amount of original uranium-238 and the final amount of stable lead-206 within that given crystal. The decay sequence from uranium to lead actually creates eight helium molecules which are then trapped within the zircon crystal. What was never measured (until recently) was the rate at which the resulting helium diffuses out of the zircon crystals.

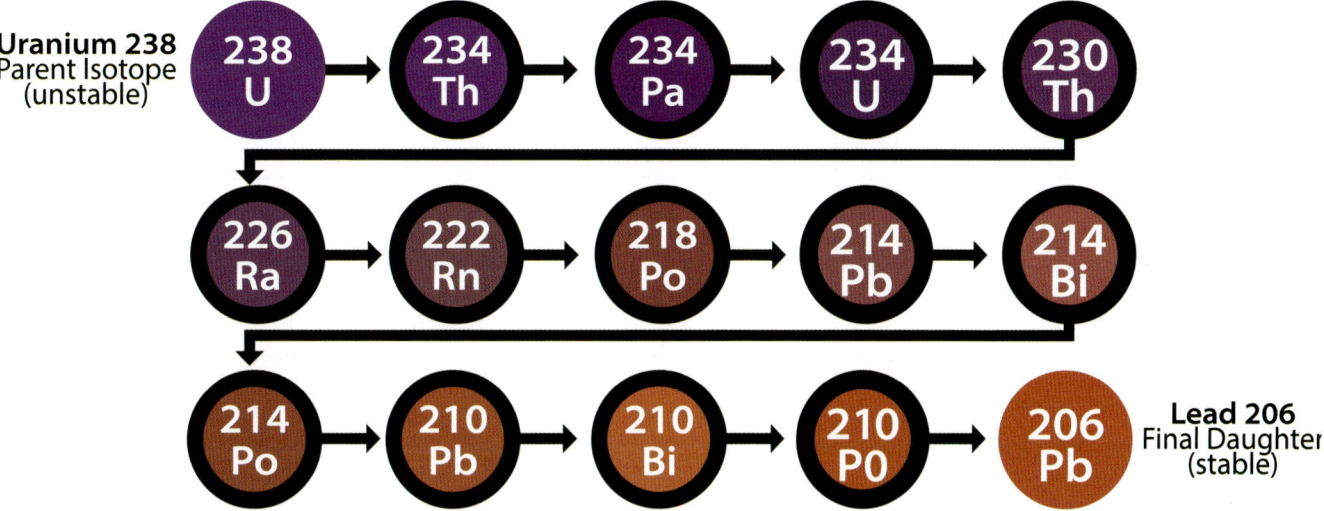

In 1974, a project by Los Alamos National Laboratories collected core samples of granite (basement rock or foundational creation rocks) which was removed from 2.6 miles deep in the earth's crust at Fenton Hill, New Mexico. The permeation rate of helium through this granite was later measured at an internationally certified laboratory. Samples of these rocks were obtained at a wide variety of depths, corresponding to different temperatures, and the permeation rate vs. temperature equation was developed. By dividing the amount of helium left in the rock with the measured permeation rate of helium through the zircon crystals, it is possible to measure how long ago the radioactive decay happened. This is the same concept as measuring the age of a helium balloon by knowing the amount of helium left in it and dividing by the rate at which the helium left the balloon. If the granite were millions of years old, there should not have been a single molecule of helium left. The result was the astounding discovery that the radiometric decay which created the helium within these zircon crystals, had left an enormous amount of the organic helium corresponding to the formation of the granite within the last 6,000 years. There is no known mechanism which could have allowed the helium to remain within these rocks for a longer period of time. However, conventional dating techniques place these rocks at over a billion years old. This is incredibly strong evidence for both the recent creation of the rock layers of our planet and the accelerated nuclear decay, which must have happened either at the time of creation (when the heavens were being "stretched out" and the earth was being supernaturally formed) or possibly during the year of the worldwide Flood.[75, 75a]

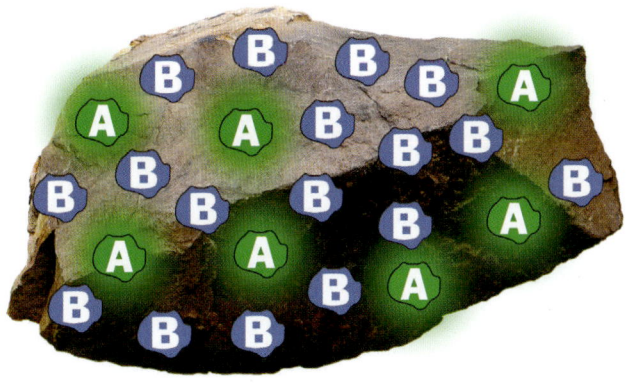

Potassium-argon dates in error

Volcanic eruption	When the rock formed	Date by radiometric dating
Mt Etna basalt, Sicily	122 BC	170,000-330,000 years old
Mt Etna basalt, Sicily	AD 1972	210,000-490,000 years old
Mt St. Helens, Washington	AD 1980	300,000-400,000 years old
Hualalai basalt, Hawaii	AD 1800 - 1801	1.44-1.76 million years old
Mt Ngaurauhoe, New Zealand	AD 1954	3.3-3.7 million years old
Kilauea Iki Basalt, Hawaii	AD 1959	1.7-15.3 million years old

This evidence for our recent creation is no different than finding a helium balloon still floating in a tomb. It is unmistakable proof that the balloon had to have been filled recently. In a similar way, the granite layers of our planet had to have also been created quite recently or the helium formed within them by radiometric decay would be long gone.

One last little known fact is that various different radiometric dating methods often give very different dates for different samples from the same layer. often give widely different dates. Even rock samples of a known age can give dates which we know are wrong. For instance, here are dates obtained from volcanic rock which were formed recently:

Evolutionists assume that these dates are wrong because the volcanic material contains material mixed with older rock. How do they know they are getting the correct date when they do not know when the rock formed? Whenever a date is obtained which does not agree with the assumed age, it is called a "discordant date" and thrown out. This method has wide variability, and researchers literally shop around for the date which agrees with what they want to accept

A Matter of Time

KEY POINTS TO REMEMBER:

1. About ½ of all radiometric decay processes produce helium, which escapes from the rock layers where it was formed.
2. Measurements of the amount of helium left show that the rock layers of the earth must have formed quite recently.
3. Radiometric dates vary widely, and known age samples often give the wrong dates.
4. Circular reasoning is used to date the age of fossils found in rock layers.

D. How do You Date Fossils in Sedimentary Rock Layers?

Most fossils are found in sedimentary rock layers which were all formed under water. These layers often have few unstable isotopes which can be used to date the rock layers. So how do we actually know how old these layers are? For instance, suppose I find an extinct animal called the trilobite in a layer of rock. How old is that layer? Evolutionists have lined up fossils since the 1800s in a tree of life and trilobites are assumed to be an ancient sea creature which evolved early in Earth's history. Assumed ages were assigned to these rock layers based on how long early researchers thought it would take for life to evolve. Much later, as it became obvious how complex life is, more time was assumed to have passed between layers. When radiometric dating was developed to date the layers below the sedimentary rock, even longer times were assigned to the various layers containing fossils. But the question remains, how do we KNOW how long ago trilobites lived? Well, what rock layer did you find it in? Since it is typically found in the lowest layer containing life, it is assumed to be the Cambrian layer. How do you know it's the Cambrian layer? Oh, I found a trilobite in it. This is circular reasoning. The rock layer dates the fossil, and the fossils date the rock layers. This is inconclusive, assumes the answer, and is lousy science.

C. Recent Creation Evidence

1. Undecayed dinosaur tissue

Dr. Mary Schweitzer, a paleontologist from North Carolina State University, discovered soft tissue of blood vessels inside fossilized T-Rex bones believed to be "68-70 million years old" from the Hell Creek formation in eastern Montana.[76] In fresh bones, any weak acid can be used to remove the rock, leaving only organic material such as fibrous connective tissue, blood vessels and various cells. By comparison, if one were to use acid to demineralize a typical fossil, nothing would be left because everything has been turned to stone. Yet, this acid-treated T. Rex bone fragment left flexible and elastic structures very similar to what one would get from a fresh bone according to Dr. Schweitzer, "so flexible and resilient that when stretched would return to its original shape."

Close examination of the broken thigh bone also revealed round microscopic structures that appear to be blood cells inside hollow vessels. Dr. Schweitzer stated, "I am quite aware that according to conventional wisdom and models of fossilization, these structures aren't supposed to be there, but there they are. I was pretty shocked." The entire evolutionary community should be far more than shocked. They should be rethinking their belief in millions of years of Earth history. How could a fossilized bone, buried under enormous heat and pressure, and after millions of years of exposure to the environment, still have intact organic structures? The soft tissue should have completely degraded. Dinosaur fossils still containing soft tissue means it is young, not millions of years old. This means the rock layers are also not millions of years old.

This is not the first soft tissue to be unearthed. Nucleic acid (DNA) with up to 800 base pairs has been taken from fossil magnolia leaves allegedly "17-20 million-years-old." Yet, DNA exposed to water and oxygen rapidly degrades and, there should not be a single base pair left after 50,000 years. A "25-40 million-year-old" bee was also found encased in amber containing living bacteria spores! Dr. Cano, the discoverer, took careful measures to avoid contamination. His analysis revealed that this bacteria's DNA is very similar to today's modern bee bacteria, Where's the evolution?

These finds -- soft tissue and DNA in fossils and amber -- are not isolated examples, but part of a growing number of scientific discoveries which are rocking the evolutionary time frame to its core! However, don't expect to hear about them in textbooks, popular science magazines, or museums. Why? Because these recent tissue samples prove that these organisms died recently. Therefore, the sedimentary rock layers of the earth were laid down recently – not millions of years ago. It is easier to deny, ignore, or explain away the implications of these finds rather than to admit that the foundational assumptions of the evolutionary belief system -- huge periods of time, the imaginary geologic column, no recent worldwide flood -- could be wrong.

More recent work by Dr. Schweitzer has shown that if iron can be attached to proteins, it slows the decay rate, and this has been widely publicized as how soft dinosaur tissue can survive so long. But do not be fooled. The experiment takes place is a sterile lab environment with anticoagulants and other chemicals not found in a natural setting. These were used to keep blood from clotting and locking up the iron molecules, as would have happened in any natural setting. The resulting increase in tissue longevity is still orders of magnitude less than needed to explain dinosaur tissue preservation. Soft tissue, undecayed proteins, and DNA fragments in fossils all testify to these rock layers having formed during Noah's Flood (about 5,000 years ago), not millions of years ago.

KEY POINTS TO REMEMBER:

1. Undecayed tissue in fossil shows that these fossils were buried recently during Noah's Flood.
2. If the fossils in the rocks are not millions of years old, the rock layers cannot be millions of years old.
3. Experiments designed to explain how iron from blood could preserve tissue for long time periods were done with sterile lab tests, and do not represent what would actually happen in an actual rock layer.

2. The amount of salt in the ocean

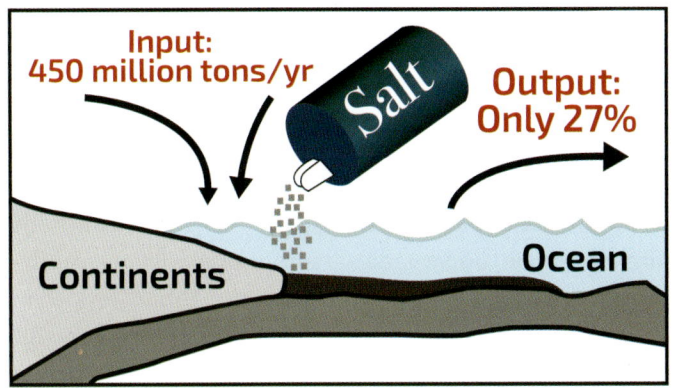

Do you realize that salt in the oceans confirms the earth is young? Every year water coming off land surfaces runs into the oceans, bringing dissolved salt with it. So, each year the oceans are becoming slightly saltier. Believe it or not, scientists have actually figured out the rates of salts going into and out of the oceans. Therefore, they have been able to calculate the maximum age of the oceans. Scientists have calculated a rate of input as 450 million tons of salt every year, with 27% leaving. When dividing the total amount of salt in the oceans by the amount flowing and remaining dissolved in the water each year, the resulting "upper age limit" for the oceans is 62 million years old. Evolutionists believe the oceans to be billions of years old. Does the salt show this? No. but you say, if the Earth is young, why is there 62 million year's worth of eroded salt in the water? Remember, there was a worldwide Flood which ground up and pulverized epic amounts of rocks releasing salts, and the fountains of the deep likely released enormous amounts of salt. So, lots of salts were added by a catastrophic worldwide Flood. God likely created the oceans with lots of initial salt. Ocean salt does not show billions of years; the ocean salt level testifies to a young earth.[77]

KEY POINTS TO REMEMBER:

1. The ocean gets saltier each year as rivers bring minerals into the ocean.
2. The Flood of Noah and initial creation of a salty ocean accounts for most of the ocean's salt.
3. The amount of salt in the ocean proves they could not be billions of years old as evolution contends.

3. Rapid decay of the Earth's magnetic field

Measurements indicate that the total energy of the earth's magnetic field has decreased by a factor of 0.7 in the last 1,000 years.[78,80] Thus 1400 years it is 1/2 as strong, i.e. Before Noah's Flood it would have been almost 10x stronger, protecting life from the suns harmful radiation much better. Without the Earth's magnetic field, life could not exist on Earth; because it is the earth's magnetic field which protects us from harmful solar radiation. The fact that Earth has a rapidly decaying magnetic field is strong evidence that we live on a young Earth, not one that is billions of years old. In spite of 100 years of attempting to explain how the earth could maintain or regenerate its own magnetic field, every attempt has failed. There is no known answer for why the Earth's magnetic field is decaying so rapidly - except for the second law of thermodynamics and the reality that the Earth has been recently created.

KEY POINTS TO REMEMBER:

1. The earth's magnetic field is rapidly decaying.
2. This magnetic field is critical for life on earth to exist because it protects us from harmful radiation from the sun.
3. The rate of decay indicates that the earth must have been recently formed.

4. Spiral galaxies

The inner stars of galaxies rotate faster than the outer stars. If spiral galaxies were more than a few hundred million years old, they should be a featureless smear instead of a distinct spiral. For 50 years this problem has been known, but no plausible explanation has been found. Even deep space pictures from NASA space telescope stun astronomers by showing fully formed spiral galaxies at the furthest edges of the universe (assumed to be 13 billion years old). Spiral galaxies say we live in a young universe. [80]

5. Rapid mutational buildup

The human genome contains at least 5,000 known genetic mistakes serious enough to cause physical deformities, diseases, cancer, mental illnesses, allergies and/or early death. With every passing generation between 100 and 1,000 random mutational mistakes are added. Fortunately, our genetic code has 3 billion "letters," and there are many error correction and backup systems to allow us to still function. However, over time, the function of every biological creature is getting worse, not better. Only 3,000 years ago, it was common and routine for close relations like siblings and cousins to marry. This happened in both the Bible and ancient civilizations because there were far fewer mistakes within the human genome to cause genetic problems. It was not until the time of Moses (3500 years ago) that so many mistakes had developed that the practice of marrying close relatives was outlawed. This rapid and unstoppable degeneration of the genetic code is strong evidence that humanity and all of creation is not millions of years old but was recently created. Evolutionary teaching states that human-like creatures have been present on earth for over a million years. How can this be? Humanity would have had a massive buildup of mutational mistakes. The deteriorating human genome clearly indicates that humans have not been on earth very long.[81]

No dating method provides an absolute date for the age of the earth. With each method, the lower limit can be zero (for instance, salt could have started out at current levels in the ocean, air may have started containing current levels of helium, etc.), but each sets a maximum possible limit. The lower limit of the age of the earth is approximately 6,000 years because this is when historical records began. All other methods are based on unprovable assumptions. Many methods give an age which agree with the Bible (less than 10,000 years) whereas 5 billion years is completely out of touch with the vast majority of ways of determining the age of the earth. Over 100 dating methods which give a recent creation date are simply ignored because it is **assumed** that God did not created mankind and therefore the Earth must be extremely old.

KEY POINTS TO REMEMBER:

1. Dinosaur soft tissue confirms a young earth.
2. The amount of salt in seawater says the earth is young.
3. Helium still in rocks is evidence for a young earth.
4. Rapid decay of earth's magnetic field indicates a young earth.
5. The existence of spiral galaxies reveals a young universe.
6. Deterioration of the human genome confirms we have not been here millions of years.

SECTION VI - CONCLUSION

Summary: Why Does it Matter?

Ideas have consequences. Truth matters, people act on what they believe to be true. If children are indoctrinated through a public education system which allows only evidence for a naturalistic explanation of life without God, we will increasingly become a society which looks only to ourselves for answers to life's questions. Less than 200 years ago, it was considered normal throughout the South Pacific for humans to eat other humans. Less than 90 years ago, Jews were declared inferior to other humans and 6 million were exterminated *"to purify the human gene pool and advance human evolution."* Over the last 100 years, over 100 million people have been murdered by communist and totalitarian nations as God had been eliminated and replaced with humanistic governments as absolute authority over right and wrong. It is not human wisdom, intellect, or evolutionary principles which saves cultures from these evils. It is accepting Jesus Christ as Savior and Lord, which results in the desire to trust the Bible and live by Biblical principles of Christian love and self-sacrifice.

Yet, if evolution is true, then the Bible is wrong about history, biology, geology, cosmology and everything else our senses reveal. So why believe it when it talks about salvation and eternity? Over 2,000 years ago Jesus made a statement relevant to this subject: *"If I have told you earthly things, and ye believe not, how shall ye believe, if I tell you of heavenly things?"* (John 3:12 – KJV)

If we have evolved from apes; if we are just another animal ...then who sets the rules? Whose standard should define right from wrong, good from bad, helpful from harmful, lawful from unlawful? If we evolved from pond slime, on what factual basis could anyone say one belief is right, while someone else's belief is wrong? Without an absolute basis for morals, the distinction between these opposites disappear. The result is a spiraling descent toward meaninglessness and a degeneration in the value of human life. Throughout history, as bacteria-to-human evolution was increasingly accepted as a fact, atheism increases. As a result, Christianity decreases, the persecution of those opposing evolution increases, and human freedom ultimately disappears. Acknowledgment of creation provides answers to foundational questions about life that are based on factual scientific and historical evidence. This topic, and your responsibility as a believer to guide the next generation to the truth, is very important indeed.

In summary, there are many MAJOR problems with evolution. These problems fall into two areas – scientific and theological. Evolution of bacteria-to-people is a theological belief which directly contradicts and undermines the only other alternative – Biblical creation. It also denies scientific observations showing its impossibility. There are many problems with promoting only evolution in schools while ignoring the evidence for creation.

- Not exposing others to the problems with evolution or showing them the evidence for creation guarantees that the majority will accept evolution.
- Not teaching the problems with evolution leaves students ill prepared to make an informed decision, with the result that most will succumb to the enormous academic pressure to accept bacteria-to-human evolution as true.
- Censorship of the creation evidence creates blindness to the truth.

A) Educational and Scientific Issues

1. The biggest problem with teaching only the evidence for cosmic, chemical, geological, and biological evolution is that none of them adequately explain the existence of the universe or the origin, complexity and diversity of life. Teaching only the evidence for evolution completely overlooks the major problems with evolution.

2. Evolution does not develop critical thinking skills. It does not explain to others how to thoughtfully and logically evaluate objective facts.

3. Teaching ONLY the evidence for molecules-to-man evolution does not allow the next generation, (who will become future teachers) the opportunity to search for the truth. **This is indoctrination, not education.**

4. The scientific method leads to three levels of certainty: hypothesis, theory, law. Molecules-to-man does not even qualify for the first level because there are so many observations which contradict it. In textbooks and in the media, evolution is taught as a fact. If evolution is a fact, why not call it the "Law of Evolution" like the Law of Gravity or the 2nd Law of Thermodynamics? Evolutionists are side stepping the scientific method by saying it is a fact. Be scientific!

5. Not teaching the flaws with evolution is not searching for the truth. Once firmly convinced that evolution is a fact, People become so indoctrinated that they cannot even consider any other option.

B) Biblical Issues

1. Evolution completely contradicts the Bible. If molecules-to-man evolution is true, then the Bible is at best irrelevant, and at worst deceiving. If bacteria-to-human evolution is true, then the Bible cannot be understood to mean what it actually says. So why accept anything it says as true or relevant?

2. Evolution puts millions of years of death, disease, bloodshed, and extinction before mankind even appeared upon the Earth. This means that if God exists at all, he created a world filled with all this death and disease, and it has always been that way. In essence, evolution makes God an evil God where the innocent have always died for no particular reason. Biblical creation acknowledges that all humans are sinful as a result of our being made in the image of God, given the freedom of choosing good over bad and obedience over disobedience; we have chosen rebellion. No one teaches children to be selfish, liars, or thieves; it comes naturally because of our fallen nature and rebellion against God. God made a perfect creation without death and death started only after the rebellion of the first humans – Adam and Eve – chose to reject the authority of their Creator. God is both perfect in justice and perfect in love. But how is God's justice and God's love reconciled? Our rebellion is what brought death into creation. Every human deserves the death sentence. But death also exists so that we will not live forever separated from our Creator. When death is viewed from this perspective, it is actually an act of mercy and love on the part of God. He did not want us to live forever and be forever separated from him. But most importantly, God reveals His incredible love and mercy taking that penalty of death upon Himself as He became a human, Jesus Christ, and died in our place.

"But God demonstrates his own love for us in this: While we were still sinners, Christ died for us." – Romans 5:8 (NIV)

3. The belief in molecules-to-man evolution means that death was in the world long before Adam sinned. If evolution were true, death is not the result of sin, and death is not the wages of sin. This would make Christ's death on the cross meaningless.

As Christians we should strive to be "teachers." Teachers, have the ability to impact and change the destiny of the next generation. All of us have beliefs based on what we have been taught by others. Once our beliefs become established, we seldom return to question our beliefs. Yet our beliefs are what drive our behavior. This is why the Bible puts a high degree of responsibility on teachers:

"...You know that we who teach will be judged more strictly."
– James 3:16(NIV)

Understanding where we came from impacts some of the most important issues of life:
- Why we exist
- Our purpose in life
- What brings meaning to life
- The nature and character of God
- What will happen after we die

Take this subject seriously and look for opportunities to teach the truth in this area. Knowing that we have a Creator lays the foundation of **knowing** that we have a Savior. It is the evidence from creation that holds us accountable for knowing that we have a Creator:

"Since what may be known about God is plain to them, because God has made it plain to them. For since the creation of the world God's invisible;e qualities - his eternal power and divine nature - have been clearly seen, being understood from what had been made, so that people are without excuse." - Romans 1:19-20 (NIV)

Because we all know that we have a Creator we stand guilty before Him for disobeying Him. The Bible also tells us that **we can know** that when we have been saved:

*"I write these things to you who believe in the name of the Son of God so that you may **know** that your have eternal life."* - 1 john 5:13 (NIV)

So how do we know that we have been saved? It is not the result of our goodness, actions, prayers, or church attendance. It is a free gift that we have not earned. As previously stated, it was while we were still sinners that God died for us, not after we "cleaned ourselves up." But we must accept it. Until we admit our sinfulness, we have no need for a Savior. God has made your salvation that God offers as the most important decision you will ever make. But understand this, IT WILL change your priorities and your life...for the better.

"For by grace you have been saved through faith, and that not of yourselves; it is the gift of God, not of works, lest anyone should boast." - Ephesians 2:8-9 (NKJV)

"Whosoever shall call on the name of the Lord shall be saved." - Acts 2:21 (KJV)

✝ ✝ ✝ ✝ ✝ ✝ ✝

✝ ✝ ✝ ✝ ✝ ✝ ✝

References and Links to Online articles

1. Bergman, Jerry, Ph.D., Journal of Creation 13(2):118–123, The attitude of various populations toward teaching creation and evolution in schools, extracted 6/10/2020
2. Doyle, Shaun, B.S., Evolution is science, but creationism is religion, https://creation.com/creation-evolution-buzzwords, extracted 6/10/20
3. Morris, Henry, Ph.D., Men of Science, Men of God: Great Scientists Who Believed in God, (1988)
4. Bergman, Jerry, Ph.D., Journal of Creation, Censorship of Information on Origins, https://creation.com/images/pdfs/tj/j10_3/j10_3_405-414.pdf, extracted 6/10/20
5. Dominic Statham, B.Sc., D.I.S., M.I.E.T., C.Eng., Strawmen and censorship: the British Humanist Association and Creation in Schools, https://creation.com/humanist-censorship, extracted 6/10/20
6. Bergman, Jerry, Ph.D., Slaughter of the Dissidents: The Shocking Truth about Killing the Careers of Darwin Doubters, (2008)
7. Bergman, Jerry, Ph.D., Journal of Creation, 23(2):37-40, "If You Can't Beat Them, Ban Them", https://creation.com/slaughter-of-the-dissidents, extracted 6/10/20
8. Batten, Don, Ph.D., Origin of Life: An explanation of What is Needed for Abiogenesis (or biopoiesis), https://creation.com/origin-of-life, extracted 6/10/20
9. Patterson, Roger, Evolution Exposed: Your Evolution Answer Book for the Classroom, pp.132-148, (2006)
10. Oard, Michael, M.S., Journal of Creation, 24(1):13-14, "Did the Early Earth's Atmosphere Contain Oxygen?", https://creation.com/did-early-earth-atmosphere-contain-oxygen, extracted 6/10/20
11. Bergman, Jerry, Ph.D., Journal of Creation 18(2):28–36, https://creation.com/why-the-miller-urey-research-argues-against-abiogenesis, extracted 6/11/20
12. Wells, Jonathan, Ph.D., Icons of Evolution Science or Myth: Why much of what we teach about evolution is wrong, pp. 9-22, (2002)
13. Grigg, Russell, M.S., Creation 13(1):30–34, "Could monkeys type the 23rd Psalm?", https://creation.com/could-monkeys-type-the-23rd-psalm, extracted 6/10/20
14. Sarfati, Jonathan, Ph.D., Journal of Creation 11(1):4–6, "Self-replicating enzymes? A critique of some current evolutionary origin-of-life models", https://creation.com/self-replicating-enzymes, extracted 6/10/20
15. Sarfati, Jonathan, Ph.D., Self-made cells? Of course not!, https://creation.com/self-made-cells-of-course-not, extracted 6/10/20
16. Sarfati, Jonathan, Ph.D., Panspermia theory burned to a crisp: bacteria couldn't survive on meteorite, https://creation.com/panspermia-theory-burned-to-a-crisp-bacteria-couldn-t-survive-on-meteorite, extracted 6/10/20
16a. Bergman, Jerry, Ph.D., Creation Ex Nihilo Technical Journal, vol. 7(1), pp. 82–87, Panspermia – The Theory that Life came from Outer Space, https://creation.com/images/pdfs/tj/j07_1/j07_1_82-87.pdf, extracted 6/10/20
17. Patterson, Roger, Evolution Exposed: Your Evolution Answer Book for the Classroom, pp.35-44, (2006)
18. Grigg, Russell, M.S., Creation 37(4):52–55, Carl Linnaeus: the scientist who saw evidence for God in everything in nature, https://creation.com/carl-linnaeus, extracted 4/18/2020
18a. Bergman, Jerry, Ph.D., Acts and Facts, Carolus Linnaeus: Founder of Modern Taxonomy, https://www.icr.org/article/carolus-linnaeus-founder-modern-taxonomy, 10/31/2014
19. Jerlström, Pierre, Ph.D., Journal of Creation 14(2):11–13, Is the evolutionary tree turning into a creationist orchard?, https://creation.com/is-the-evolutionary-tree-changing-into-a-creationist-orchard, extracted 6/10/20
20. Statham, Dominic, B.S., Creation 34(4):43–45, Homology made simple, https://creation.com/homology-made-simple, extracted 6/10/20
21. Bergman, Jerry, Ph.D., Journal of Creation 15(1):26–33, Does homology provide evidence of evolutionary naturalism?, https://creation.com/does-homology-provide-evidence-of-evolutionary-naturalism, extracted 6/10/20
22. Oard, Michael, M.S., Journal of Creation 25(2):22–31, Did birds evolve from dinosaurs?, https://creation.com/bird-evolution, extracted 6/10/20
23. Carter, Robert, Ph.D., Did Dinosaurs evolve into birds?, https://creation.com/dinosaur-bird-evolution, extracted 6/10/20
24. Tomkins, Jeffery, Ph.D., Journal of Creation 26(1):94–100, Genomic monkey business—estimates of nearly identical human–chimp DNA similarity re-evaluated using omitted data, https://creation.com/human-chimp-dna-similarity-re-evaluated, extracted 6/10/20
24a. Catchpoole, David, Ph.D., Creation 33(2):56, Y chromosome shock, https://creation.com/y-chromosome-shock, extracted 6/10/20

24a. Batten, Don, Ph.D., Creation 36(1):35–37, The myth of 1%, https://creation.com/1-percent-myth, extracted 6/10/20

25. Anderson, Daniel, Petrified Wood: Fast or Slow?, https://creation.com/petrified-wood-fast-or-slow, Extracted 6/10/20

26. Snelling, Andrew, Ph.D. Creation 17(4):38–40, 'Instant' petrified wood, https://creation.com/instant-petrified-wood, extracted 6/10/20; Creation, "Wood Petrified in Spring", June-August 2006, 18-19, extracted 6/10/20

27. Robinson, Steven, Ph.D., Creation Ex Nihilo Technical Journal 10(1), Can Flood Geology Explain the Fossil Record?, https://creation.com/images/pdfs/tj/j10_1/j10_1_032-069.pdf, extracted 6/10/20

27a. Morris, John D. and Frank J. Sherwin, Ph.d., The Fossil Record Unearthing Nature's History of Life. ICR, 2010, p.41.

28. Dickens, Harry, Ph.D., Journal of Creation 30(1):8–10, The 'Great Unconformity' and associated geochemical evidence for Noahic Flood erosion, https://creation.com/evidence-for-noahic-flood-erosion, extracted 6/10/20

29. Wieland, Carl, Ph.D., Creation 16(2):38–39, Exploding evolution, https://creation.com/exploding-evolution, extracted 6/10/20.

30. Wieland, Carl, Ph.D., Creation 19(3):40–43, Probing the earth's deep places, https://creation.com/probing-the-earths-deep-places, extracted 6/10/20

31. Baumgardener, John, Ph.D., Journal of Creation 16(1):58–63, Catastrophic plate tectonics: the geophysical context of the Genesis Flood, https://creation.com/catastrophic-plate-tectonics-the-geophysical-context-of-the-genesis-flood, extracted 6/10/2020

32. Catchpoole, David, Ph.D., 'Living fossils' enigma, https://creation.com/living-fossils-enigma, extracted 6/10/20

33. Batten, Don, Ph.D., Creation 33(2):20–23, Living fossils: a powerful argument for creation, https://creation.com/werner-living-fossils, extracted 6/10/20

34. Werner, Carl, Ph.D, Living Fossils – Evolution: The Grand Experiment, pp.24-224 (2008)

35. Oard, Michael, M.S., Journal of Creation 23(2):17–24, How old is the Grand Canyon?, https://creation.com/how-old-is-grand-canyon, extracted 6/10/20

35a. https://answersingenesis.org/geology/grand-canyon-facts/when-and-how-did-the-grand-canyon-form/, retrieved 7/7/2020

35b. https://answersingenesis.org/geology/grand-canyon-facts/when-and-how-did-the-grand-canyon-form/, retrieved 7/7/2020

36. Sarfati, Johnathan, PhD., Did a meteor wipe out the dinosaurs?, https://creation.com/did-a-meteor-wipe-out-the-dinosaurs-what-about-the-iridium-layer, extracted 6/10/20

37. Oard, Michael, M.S., Journal of Creation 11(2):137–154, Extinction of the Dinosaurs, https://creation.com/the-extinction-of-the-dinosaurs, extracted 6/10/20

38. Sanford, John, Ph.D., Genetic Entropy: The Mystery of the Genome, (2005)

39. Carter, Rob, Ph.D., Genetic entropy and simple organisms, https://creation.com/genetic-entropy-and-simple-organisms, extracted 6/10/20

40. Von Vett, Julie; Malone, Bruce, Inspired Evidence: Only One Reality, p. 7/9, (2011), https://creation101.org

41. Wieland, Carl, Ph.D., Creation 14(3):22–23, Darwin's Finches, https://creation.com/darwins-finches, extracted, 6/10/20

42. Von Vett, Julie, Malone, Bruce, Have You Considered: Evidence Beyond a Reasonable Doubt, p. 4/22, (2017)

43. Cody J. Guitard, Cody, M.S., Journal of Creation 32(2):20–28, Creationist modelling of the origins of Canis lupus familiaris—ancestry, timing, and biogeography, https://creation.com/dog-origins, extracted, 6/10/20

44. Wieland, Carl, Ph.D., Creation 21(3):56, Goodbye, peppered moths, https://creation.com/goodbye-peppered-moths, extracted 6/10/20

45. Malone, Bruce, Search for the Truth: Changing the World with the Evidence for Creation, p. 23, (2011)

46. Batten, Don, Ph.D., Creation 39(4):46–48, Antibiotic resistance: Evolution in action?, https://creation.com/antibiotic-resistance-not-evolution-in-action, extracted 6/10/20

47. Biswas, Chinmoy, Ph.D., Journal of Creation 20(2), Founder Mutations: Evidence for Evolution?, https://creation.com/images/pdfs/tj/j20_2/j20_2_16-17.pdf, extracted 6/10/20

48. Wieland, Carl, Ph.D., New eyes for blind cave fish?, https://creation.com/new-eyes-for-blind-cave-fish, extracted 6/10/20

49. Batten, Don, Ph.D., Bacteria 'evolving' in the lab?, https://creation.com/bacteria-evolving-in-the-lab-lenski-citrate-digesting-e-coli, extracted April 2020.

50. Bergman, Jerry, Ph.D., Useless Organs: The Rise and Fall of a Central Claim of Evolution, (2019)
51. Bergman, Jerry, Ph.D., Journal of Creation 14(2):95–98, Do any vestigial organs exist in humans?, https://creation.com/do-any-vestigial-organs-exist-in-humans, extracted 6/10/20
52. Statham, Dominic, B.S., More nails in the coffin of 'junk DNA', https://creation.com/junk-dna-functions, extracted 6/10/20
53. Walkup, Linda, Ph.D., Journal of Creation 14(2):18–30, 'Junk' DNA: evolutionary discards or God's tools?, https://creation.com/junk-dna-evolutionary-discards-or-gods-tools, extracted 6/10/20
54. E. van Niekerk, E. van, M.S., Journal of Creation 25(3):89–95, Ernst Haeckel, fraud is proven, https://creation.com/haeckel-fraud-proven, extracted 6/10/20
55. Parker, Gary, Ph.D., Creation 6(3):6–9, Creative design in the human embryo, https://creation.com/embryo-design, extracted 6/10/20
56. Sutherland, Luther, Darwin's Enigma, pp. 88 – 90, (1998)
57. Sarfati, Jonathan, Ph.D., Creation 21(3):28–31, The non-evolution of the horse, https://creation.com/the-non-evolution-of-the-horse, extracted 6/10/20
58. Safarti, Jonathan, Ph.D., Refuting Evolution: A handbook for students, parents, and teachers countering the latest arguments for evolution – Chapter 5 'Whale Evolution', https://creation.com/refuting-evolution-chapter-5-whale-evolution. Extracted 6/10/20
59. Sutherland, Luther, Darwin's Enigma, p. 122, (1998)
60. Gould, Stephen, Ph.D.,"The Return of the Hopeful Monster", Natural History, Vol. LXXXVI(6), p.24, (1977)
61. Kitts, David, Ph.D.,"Paleontology and the Evolutionary Theory", Evolution, vol. 28, p.467, (1974)
62. Ridley, Mark, Ph.D., "Who Doubts Evolution?", New Scientist, Vol. 90, No. 1259, pp. 830-832, (1981)
63. Stanley, Steven, Ph.D., Macroevolution: Pattern and Process, p. 39, (1979)
64. Oard, Michael, M.S., Creation 25(4):10–14, Neanderthal Man–the changing picture, https://creation.com/neanderthal-manthe-changing-picture, extracted 6/10/20
65. Creation 13(3):22–23, Who was Java Man, https://creation.com/who-was-java-man, exported 6/10/20
66. Mehlert, Bill, Creation Ex Nihilo Technical Journal, vol. 8(1), pp. 105–117, Homo erectus 'to' Modern Man:Evolution or Human Variability?, https://creation.com/images/pdfs/tj/j08_1/j08_1_105-116.pdf, extracted 6/10/20
67. Rupe, Christopher, M.S.; Sanford, Jon, Ph.D., Contested Bones, pp.27-246 (2017)
67a. Wieland, Carl. Ph.D., Lucy: walking tall–or wandering in circles?, https://creation.com/lucy-walking-tall-or-wandering-in-circles, extracted 6/10/20
68. Grigg, Russell, M.S., Creation 25(1):16–19, Are there apemen in your ancestry, https://creation.com/are-there-apemen-in-your-ancestry, extracted 6/10/20
69. Psarris, Spike, M.S., The secular dilemma, https://www.creationastronomy.com/the-secular-dilemma/, creationastronomy.com, extracted 6/10/20
70. Bates, Gary, Multiverse theory—unknown science or illogical raison d'être?, https://creation.com/multiverse-theory, extracted 6/10/20
71. Hartnett, John, Ph.D., Creation 37(3):48–51, Big bang beliefs: busted, https://creation.com/big-bang-beliefs-busted, extracted 6/10/20
72. Hartnett, John, Ph.D., Stars just don't form naturally–'dark matter' the 'god of the gaps' is needed, https://creation.com/stars-dont-form-naturally, extracted 6/10/20
73. DeYoung, Donald, Ph.D., Thousands...Not Billions: Challenging the Icon of Evolution, Questioning the Age of the Earth, pp. 45-64, (2005)
74. ibid, pp. 65-80, (2005)
75. Morris, John, Ph.D., The Young Earth: The Real History of the Earth – Past, Present, and Future, pp. 42-119, (2005)
75a. https://answersingenesis.org/age-of-the-earth/6-helium-in-radioactive-rocks/
76. Schweitzer, Mary, Ph.D., et al.,Science, 307, no. 5717, pp. 1952-1955, (2005)
77. Morris, John, Ph.D., The Young Earth: The Real History of the Earth – Past, Present, and Future, pp. 86-87, (2005)
78. Morris, John, Ph.D., The Young Earth: The Real History of the Earth – Past, Present, and Future, pp. 76-83, (2005)
79. Howells, Joshua, Spiral Galaxies: To many for Big Bang, www.creation.com
80. Sanford, John, Ph.D., Genetic Entropy, (2014)

DEVOTIONAL SPECIAL
GET ALL FOUR DEVOTIONALS AT ONE LOW PRICE !

Read one and give one away
4 FOR $35.00

Inspired Evidence

This extensively illustrated devotional provides a daily reminder that the truth of the Bible is all around us. Arranged in an enjoyable devotional format, this 432 page book starts each day with yet another reason to trust God's word. There is no conflict between science and the Bible – share that truth with others.

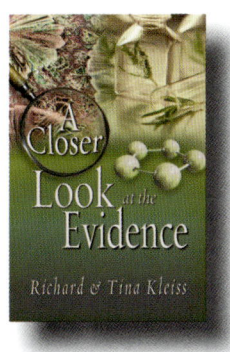

A Closer Look at the Evidence

This book offers unique evidence, primarily scientific, for the existence of our Creator. It is organized into 26 different subject areas and draws from over 50 expert sources. Each page highlights awe-inspiring examples of God's incredible handiwork.

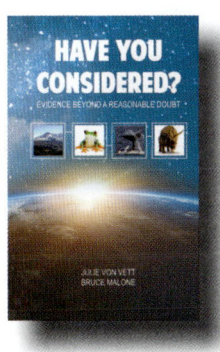

Have You Considered

This hardcover book is the third in a series of unique devotions presenting evidence for every day of the year that demonstrates it is beyond ANY reasonable doubt that we have a Creator.

This book is written in a style that anyone will understand and appreciate; extensively illustrated and meticulously documented.

Without Excuse

Our fourth creation devotional with scientific evidence supporting Biblical truth for every day of the year. A great resource for the entire family.

OTHER CREATION BOOKS
BY SEARCH FOR THE TRUTH MINISTRIES

Brilliant
This extensively illustrated devotional provides a daily reminder that the truth of the Bible is all around us. Arranged in an enjoyable devotional format, this 432 page book starts each day with yet another reason to trust God's word. There is no conflict between science and the Bible – share that truth with others.

Censored Science
This book offers unique evidence, primarily scientific, for the existence of our Creator. It is organized into 26 different subject areas and draws from over 50 expert sources. Each page highlights awe-inspiring examples of God's incredible handiwork.

A Christian's Guide to Refuting Evolution
This full color, extensively illustrated book is filled with activities, video links, and demonstrations. The book systematically reveals that the "best" evidence for evolution is riddled with contradictions and misconceptions. (8" x 11", 112 pages)

Science Shadowlands
40 short intriguing reflections on the evidence for creation that blends poetry, history, and everyday observations with science. A TOTALLY unique look at the evidence for creation. (5.5" x 8.5", 128 pages)

THE ROCKS CRY OUT CURRICULUM:

- 19 creation science presentations (approximately 50 minutes each)
- A video presentation of the creation science assembly given in the overseas schools
- A documentary of the impact these presentations and books have on students
- 60 short video creation teachings (approximately 3 minutes each

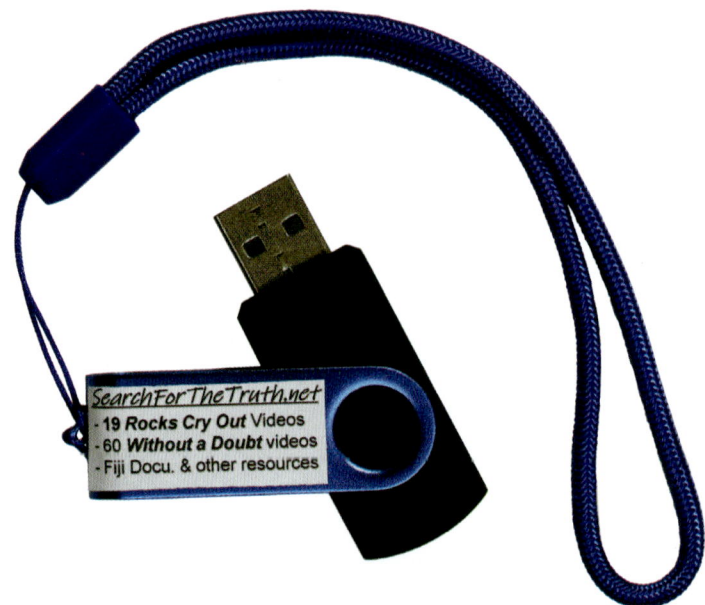

Bring the most visual, interactive, and relevant series on the evidence to creation to your church, fellowship, or youth group! Filmed at locations across America with video illustrations and animations, these lessons are not a boring technical lecture.

These 45 minute classes enable the non-scientist to bring the evidence for biblical creation to their home or church. This curriculum uses short, personal narrative-style teachings to connect God's Word with science and history, i.e "the real world". Leaders guide included with each set.

Perfect for small group, home school, or Sunday school groups of all ages, The Rocks Cry Out show how EVERY area of science confirms Biblical Truth.

SEE ALL OF OUR RESOURCES AT WWW.SEARCHFORTHETRUTH.NET

SEARCH FOR THE TRUTH
MAIL-IN ORDER FORM
See more at www.searchforthetruth.net

Call us, or send this completed order form (other side of page) with check or money order to:

Search for the Truth Ministries
3255 Monroe Rd.
Midland, MI 48642
989.837.5546 or truth@searchforthetruth.net

PRICES

	Item Price	2 - 9 Copies	10 Copies	Case Price
DEVOTIONAL SPECIAL (4 books)	$45.00	-	Mix or Match	-
Have You Considered (Hardback)	$13.95	$8.96/ea.	$8.00/ea.	call
Explore the World (Hardcover)	$13.95	$11.95/ea.	$8.00/ea.	call
A Closer Look at the Evidence (Hardback)	$13.95	$8.96/ea.	$8.00/ea.	call
Without Excuse (Hardback)	$13.95	$8.96/ea.	$8.00/ea.	call
Censored Science (Hardback)	$16.95	$11.95/ea.	$8.00/ea.	call
Brilliant (Hardback)	$16.95	$11.95/ea.	$8.00/ea.	call
Search for the Truth (book)	$11.95	$8.96/ea.	$6.00/ea.	call
Christian's Guide to Refuting Evolution (Softcover)	$12.95	$9.95/ea.	$7.00/ea.	call
Science Shadowlands (Softcover)	$11.95	$8.96/ea.	$6.00/ea.	call
Inspired Evidence (Softback)	$11.95	$8.96/ea.	$6.00/ea	call
Rocks cry out (Flash drive)	$45	-	-	call

MAIL-IN ORDER FORM

RESOURCE	Quantity	Cost each	Total
DEVOTIONAL SPECIAL (4 books)			
Explore the World (Hardback)			
A Closer Look at the Evidence (Hardback)			
Have You Considered (Hardback)			
Without Excuse (Hardback)			
Censored Science (Hardback)			
Brilliant (Hardback)			
Search for the Truth (book)			
Inspired Evidence (Softcover Book)			
Christian's Guide to Refuting Evolution (Softcover)			
Science Shadowlands (Softcover)			
Rocks Cry Out (Flash Drive)			
Tax deductible Donation to ministry			
		Subtotal	
		MI residents add 6% sales tax	
		Shipping add 20% of subtotal	
		TOTAL ENCLOSED	

Normal delivery time is 1-2 weeks

For express delivery increase shipping to 25%

SHIP TO:

Name: _____

Address: _____

City: _____

State: _____ Zip: _____

Phone: _____

E-mail: _____